BUSINESS CONDITIONS IN MICHIGAN METROPOLITAN AREAS

By
Paul J. Kozlowski

with assistance
from
Phyllis R. Buskirk

December 1979

THE W. E. UPJOHN INSTITUTE FOR EMPLOYMENT RESEARCH

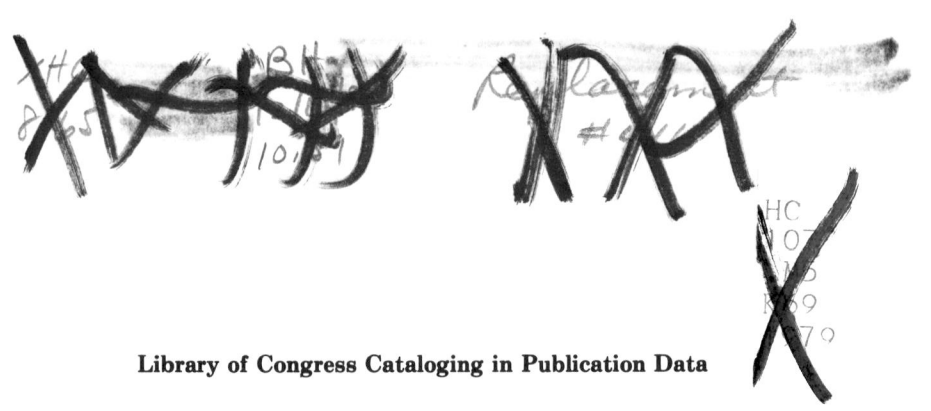

Library of Congress Cataloging in Publication Data

Kozlowski, Paul J
  Business conditions in Michigan metropolitan areas.

  1. Michigan—Economic conditions. 2. Metropolitan areas— Michigan. I. Title.
  HC107.M5K69   330.9'774'04   79-24777
  ISBN 0-911558-71-3

Copyright © 1979
by the
W.E. UPJOHN INSTITUTE
FOR EMPLOYMENT RESEARCH

300 South Westnedge Ave.
Kalamazoo, Michigan 49007

THE INSTITUTE, a nonprofit research organization, was established on July 1, 1945. It is an activity of the W. E. Upjohn Unemployment Trustee Corporation, which was formed in 1932 to administer a fund set aside by the late Dr. W. E. Upjohn for the purpose of carrying on "research into the causes and effects of unemployment and measures for the alleviation of unemployment."

This study is for sale by the Institute at $4.50 per copy. For quantity orders of this publication or any combination of Institute publications, price reductions are as follows: 10-25 copies, 10 percent; 26-50, 15; 51-100, 20; over 100, 25.

The Board of Trustees of the W. E. Upjohn Unemployment Trustee Corporation

Preston S. Parish, Chairman
Mrs. Ray T. Parfet, Vice Chairman
Charles C. Gibbons, Vice Chairman
D. Gordon Knapp, Secretary-Treasurer
E. Gifford Upjohn, M.D.
Donald S. Gilmore
Mrs. Genevieve U. Gilmore
James H. Duncan
John T. Bernhard

The Staff of the Institute

E. Earl Wright, Director
Saul J. Blaustein
Judith K. Brawer
Phyllis R. Buskirk
H. Allan Hunt
John R. Mekemson
Philip M. Scherer
Robert A. Straits
Jack R. Woods

Foreword

Each of the metropolitan areas in Michigan exhibits economic characteristics that affect business conditions in that locale. Because of differences in economic structure, each area also experiences varying economic behavior in response to business cycles. In this study, fourteen comparable time series have been used to analyze business conditions in each area in terms of labor market, construction, and banking activity over a ten-year period from the late 1960s to the late 1970s. In addition, elements of the economic structure of the eleven metropolitan areas are outlined.

The Institute is pleased to present this publication as an addition to the information available regarding Michigan and the metropolitan areas and as an extension of business conditions analysis by use of local indicators for substate areas.

Facts and observations expressed in this study are the sole responsibility of the authors. The viewpoints that are presented do not necessarily represent positions of the W. E. Upjohn Institute for Employment Research.

E. Earl Wright
Director

Kalamazoo, Michigan
December 1979

Contents

Foreword v

Introduction 1
    The National Setting 3
    Regional Cyclical Sensitivity 6
    Michigan Metropolitan Areas 8

PART I   METHODOLOGY AND COMPARATIVE
             ANALYSIS OF ECONOMIC INDICATORS 11

    Measures of Local Business Activity
    and Intrastate Variations 13
        Local Time Series 14
           - Labor Market Data 14
           - Construction Data 16
           - Banking Data 17

        Cyclical Behavior 18
           - Timing 19
           - Amplitude 25

PART II MICHIGAN METROPOLITAN AREAS 35
    Ann Arbor-Ypsilanti SMSA 37
    Battle Creek SMSA 51
    Bay City SMSA 65
    Detroit SMSA 77
    Flint SMSA 93
    Grand Rapids SMSA 107
    Jackson SMSA 121
    Kalamazoo-Portage SMSA 133
    Lansing-East Lansing SMSA 147
    Muskegon-Norton Shores-Muskegon Heights SMSA . 159
    Saginaw SMSA 171

Introduction

This study focuses on economic activity in Michigan metropolitan areas over roughly a ten-year period, from the late 1960s to the late 1970s. A time marked by a fairly high degree of cyclical instability, stubborn inflationary pressures, an energy crisis, and an unsettled international environment, this period has been a particularly trying one for many of the older, industrialized urban areas located in the Northeast and North Central regions. In fact, as short-run business activity in those areas tended to conform closely to national business cycles, the longer term problems associated with urban decay and structural adjustments have been magnified in some cases by the instability of the national economy. Also, shifts in emphasis toward income maintenance programs designed to meet immediate hardships associated with job losses during business slumps have tended to take precedence over urban revitalization and industrial expansion programs. As a result, a local slump has often meant that economic redevelopment efforts, limited to some degree by economic conditions, have been slow to regain momentum. While the main concern of the study is to provide an overview of the effects of cyclical swings, growth rates for the period are compared as well.

Because of its heavy dependence on the volatile automobile industry, Michigan has a long history as a cyclically sensitive state. Over the ten-year period, the state's widely recognized boom and bust cycle has been repeated. Important questions addressed in this study are: What actually happened in Michigan metropolitan areas during this unsettled period? How does local behavior compare with the statewide performance? Which metropolitan areas exhibited the highest degree of cyclical sensitivity? And, which local economic activities appear to be most sensitive to swings in national business activity? Answers to these questions are likely to help business

2   Introduction

assess the impacts of business slumps in particular market areas and may help state and local officials formulate appropriate policy. In addition, this study reveals strengths and weaknesses in the local economies, and the results can be used to assess the probable impacts of national business cycles on these areas in the future.

In this study short-run business conditions in Michigan metropolitan areas are analyzed by measuring the behavior of a set of economic indicators covering three major categories: labor market, construction, and banking activity. Local fluctuations in overall business activity are likely to be manifested in numerous processes that occur in each of those categories. The underlying assumption in the broad approach taken here is that, like national business cycles, local cyclical swings are widely diffused across numerous economic processes, and the timing and magnitude of these fluctuations vary. From this perspective, single indicators, like total employment or the unemployment rate, while certainly useful, provide only a partial view of the performance of a local economy. A more detailed examination of local behavior, covering a wider range of activities, is regarded as much more helpful for both private and public decision making and for implementing policy at an appropriate time. Furthermore, the behavior of a particular activity reflected in a local time series, the unemployment rate, for example, is likely to exhibit similarities and differences among areas depending on differences in industrial composition and economic growth, among other things. However, identifying the causes of variations in economic performance among Michigan metropolitan areas is beyond the scope of this study. The more modest objective here is to measure and compare intermetropolitan performance in Michigan across comparable economic processes.

The study is organized into two parts. Part I discusses the data and methods used to measure local business activity and presents a comparative analysis of behavior in Michigan metropolitan areas. Intrastate differences in the timing, duration, and amplitude of cyclical swings in numerous economic indicators are emphasized. In cases where comparable national and state series exist, measures of their behavior are used as benchmarks for assessing local performance. Part II presents detailed descriptions of what happened in each metropolitan area. The eleven sections—one for each metropolitan area in the state—include a brief introductory discussion of the area's economic structure and a detailed look at local business conditions reflected in the behavior of time series covering labor market, construction, and banking activity. Charts for the local series are presented in each section, with the organization of material patterned closely after the Upjohn Institute's *Business Conditions in the Kalamazoo Area: A Quarterly Review*, which has been published since 1958.[1] For those interested in a particular metropolitan area, the individual sections in Part II provide a detailed economic profile.

Although this study concentrates on the recent historical performance in Michigan metropolitan areas through 1978, the set of local indicators shown in the sections in Part II provides a foundation for current business conditions analysis. In fact, one of the major results of this study is the presentation of one use of local time series for current business conditions analysis at the substate level.

The National Setting

It is important at the outset to have a clear picture of the nation's overall economic performance, because that determines to a considerable degree the general business environment for states and local areas located in the highly industrialized East North Central region of the country. Notable features of the last ten years include the two recessions and stubborn inflationary pressures. During that period the nation experienced nine quarters, or slightly more than two years, of slumping business activity resulting in declines in output, employment, and inflation, along with increases in unemployment. In the subsequent recovery-expansion phase, the situation was reversed following the usual business cycle pattern.

However, there were marked differences in the severity of the two recessions. The first business slump began in late 1969 after nearly nine years of economic expansion. That recession ended a year later.[2] By historical standards this business slump was mild; in fact,

---

1. There is one major difference between the presentation in Part II and *Business Conditions in the Kalamazoo Area*. In the latter, particularly in the lead article, local economic activity is divided into four categories: industrial activity, labor market conditions, banking activity, and construction. The first category, industrial activity, is based on the Institute's quarterly survey of major manufacturing firms in the Kalamazoo area. That type of information is not available for other metropolitan areas in Michigan. For an example of how such survey data are used for local business conditions analysis, see Paul J. Kozlowski, "Review of Economic Activity," *Business Conditions in the Kalamazoo Area: A Quarterly Review*, January, April, July, or October, 1978 and January, April or July, 1979.

2. Widely accepted business cycle dates have been designated by the National Bureau of Economic Research, Inc. These dates represent points in time when aggregate economic activity reached a cyclical peak or trough. This reference

4   Introduction

CHART 1
REAL GROSS NATIONAL PRODUCT
PERCENT CHANGE AT ANNUAL RATE

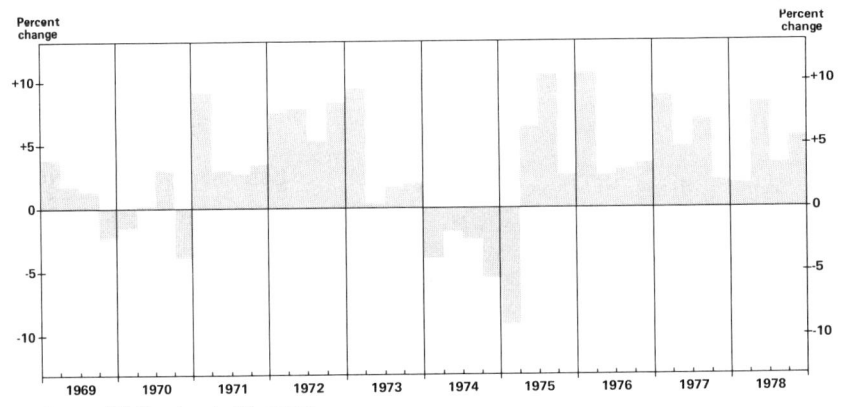

Source: U.S. Department of Commerce.

CHART 2
UNEMPLOYMENT RATE AND
GNP IMPLICIT PRICE DEFLATOR

(Seasonally Adjusted)

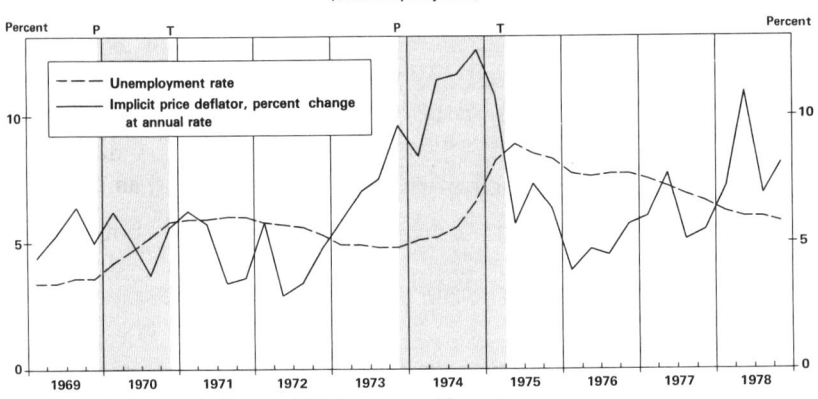

- - - - Unemployment rate
——— Implicit price deflator, percent change at annual rate

Source: U.S. Department of Labor and U.S. Department of Commerce.
Note: Shaded areas indicate national recession periods as defined by the National Bureau of Economic Research, Inc. P = peak and T = trough.

it was the mildest of the six recessions which have occurred in the United States since the late 1940s. During the 1969-70 recession, real Gross National Product (GNP) declined 1.1 percent from peak to trough, nonagricultural employment fell about 1.0 percent, and the unemployment rate rose from 3.4 percent to 6.0 percent. By comparison, the recession that began in late 1973 and ended in early 1975 was considerably more severe. Chart 1 shows the annualized percent changes in real GNP by quarter for the 1968-78 period. Note that negative growth of real GNP occurred throughout the 1973-75 recession and was of much larger magnitude than that of the 1969-70 recession. Overall, real GNP fell 6.6 percent and nonagricultural employment dropped 2.1 percent from peak to trough, and the aggregate unemployment rate jumped from a cyclical low of 4.8 percent to a high of 8.9 percent (see Chart 2). Furthermore, the 1973-75 recession was also characterized by a severe slump in construction activity, manifested in a plunge in new private housing starts from an annual rate of about 2.4 million in the fourth quarter of 1972 to slightly less than one million in the first quarter of 1975.

Output and employment rose during the business cycle expansions following each recession and, as Chart 2 shows, the unemployment rate declined. However, the nation was plagued by inflationary pressures which rose substantially when business activity picked up. Chart 2 reveals that the declines in the inflation rate during the recessions failed to offset these upswings, despite the fact that the 1973-75 slump was severe and extremely costly in terms of lost output and income. Consequently, the nationwide inflation rate exhibits an upward trend for the period as a whole.

However, the inflation rate was influenced by events not linked to swings in domestic business activity. In 1971, the Nixon Administration imposed price controls which checked the upswing in inflation early in the business expansion. When controls were phased out, the inflation rate again moved upward, exacerbated by the Arab oil embargo of 1973 and cartel pricing by the Organization of Petroleum Exporting Countries (OPEC), which pushed oil prices even higher. As Chart 2 shows, the inflation rate hit a peak of over 12 percent in late 1974. The cyclical decline in the inflation rate was

---

chronology, therefore, delineates phases of the national business cycle according to classical definitions and is used widely as a benchmark for measuring the behavior of specific national and subnational activities. Dates for the 1969-70 recession are: December 1969—peak, and November 1970—trough; and for the 1973-75 recession: November 1973—peak, and March 1975—trough. The shaded areas in Chart 2 (viii) and in the charts in Part II represent these national recession periods.

steep, and by early 1976 it had hit a low of about 3.9 percent. After that time, the inflation rate experienced another upswing as the economy settled into its most recent business cycle expansion phase.

By the end of 1978 the economy was still in an expansion phase that began in early 1975. Output, employment, and income rose substantially during the expansion, and the unemployment rate fell steadily. Nevertheless, as Chart 2 shows, the jobless rate at the end of 1978 was still above its pre-recession low and exhibited an upward drift over the ten-year period. Thus, the period examined in this study was marked by a secular rise in both the inflation and unemployment rates. By historical standards, the recent business expansion, which has been ongoing for more than four years, is "old" for peacetime periods, and in early 1979 sporadic signals of an impending recession appeared. Fueled by rising energy prices, the inflation rate has again moved above 10 percent. With personal income growth failing to keep pace, consumer purchasing power has waned, setting the stage for a falloff in aggregate demand. The expected slowdown in business activity is again likely to be more severe in those areas that are particularly sensitive to swings in aggregate spending. The state of Michigan ranks first in this respect.

Regional Cyclical Sensitivity

The heavily industrialized East North Central region, which includes Ohio, Indiana, Illinois, Wisconsin, and Michigan, is highly responsive to swings in national business activity. In fact, the five-state region ranks first among the nine Census divisions in terms of the cyclical sensitivity of employment, according to a recent study published by the Federal Reserve Bank of Boston.[3] The region had an average trend-adjusted decline in employment of 139.9 percent during the last six recessions relative to the national employment change. What this means is that employment contractions for the five-state region exceeded the national decline by 39.9 percent on the average. During expansion periods the region's employment rose an average of 42.2 percent more than nationwide.[4] These relative amplitude measures for the East North Central region are

---

3. Richard F. Syron, "Regional Experience During Business Cycles—Are We Becoming More or Less Alike?" *New England Economic Review*, November/December 1978, pp. 25-34.

4. Syron, pp. 30-31.

substantially above those of the other eight Census divisions, leading to the following conclusion: "The East North Central region with its high proportion of durable goods manufacturing is the most cyclical part of the country by any measure of employment change."[5]

The five states in the East North Central region are far from homogeneous, however. Michigan, with its heavy dependence on the automobile industry, is the most cyclically sensitive state in the region and, as already mentioned, has a long standing reputation as such because of its past boom and bust performance. A recent example is the performance of nonfarm income in the state during the 1973-75 recession and in the early part of the recovery. During the slump from the fourth quarter of 1973 to the first quarter of 1975, nonfarm income in Michigan rose only 4.6 percent in current dollars, the smallest rise among all the states.[6] That slow income growth was due largely to the 15.3 percent drop in wage and salary income in the manufacturing sector, the largest decline among the states. However, one year after the recession ended, manufacturing income in Michigan was up 23.1 percent, and total nonfarm income 12.1 percent, the latter exceeded only by the 18.6 percent rise in Alaska during the first year of recovery. This sharp turnaround in Michigan reflects the behavior of the manufacturing sector, particularly the automobile industry.

Industry mix, narrowly defined as the degree of concentration in durable goods manufacturing, appears to be one of the most important determinants of regional cyclical responsiveness, although it is certainly not the sole factor accounting for differences in regional performance. In a recent study of the effects of industry mix on regional fluctuations in income over the 1958-76 period, Lynn Browne concludes that "Industry mix is one of the factors responsible for regional differences in cyclical behavior. It is not the sole factor; nor is it always equally important."[7] Browne goes on to point out that, "the East North Central division and New England, both with large shares of income derived from durable goods manufacturing, are more exposed to short-term fluctuations than is the country as a whole."[8]

---

5. Syron, p. 31.
6. Robert B. Bretzfelder, "Contrasting Developments in the States During Recession and Early Recovery," *Survey of Current Business*, April 1976, p. 32.
7. Lynn E. Browne, "Regional Industry Mix and the Business Cycle," *New England Economic Review*, November/December 1977, p. 48.
8. Browne, p. 48.

8   Introduction

While economic growth and the degree of industrial diversification have also been linked to variations in regional cyclical response, industry mix stands out as perhaps the key factor; and it is a useful, although admittedly rough and inconsistent, predictor of the sensitivity of regions to national business cycles. Certainly, Michigan's relatively large cyclical swings in employment and income are attributable to the volatility of its industrial base. But since no two business cycles (or individual phases for that matter) are exactly the same, and since a region's industrial structure changes over time, variations in timing and amplitude of fluctuations in the same economic process can be expected from one business cycle to the next, depending upon the way in which fluctuations in the components of aggregate demand—consumption, investment, government, and net export spending—impact on and filter through that structure and how that structure has changed.

Michigan Metropolitan Areas

In 1977 Michigan's population was estimated at 9.1 million. About 80 percent of Michigan residents lived in the 11 metropolitan areas shown in Table 1. All are located in the southern half of Michigan's lower peninsula, and they stretch contiguously from the Detroit SMSA in the east to the Muskegon-Norton Shores-Muskegon Heights SMSA in the west.[9]

These 11 metropolitan areas can be viewed as small, open, regional economies which vary in size, industrial structure, and rates of growth. The six-county Detroit SMSA had a 1977 population of 4,370,200, more than seven times the size of the Grand Rapids SMSA, which ranked second. Table 1 shows that, except for the Detroit SMSA, the metropolitan areas are mostly small-population areas. Outside of Detroit, only the Grand Rapids and Flint SMSAs had population totals above one-half million. Note that the smallest metropolitan area in the state, Bay City, had a 1977 population of 120,200, or one-thirty sixth the size of the Detroit SMSA.

The state's industrial activity is also largely concentrated in these areas. In 1977, for example, 86.6 percent of manufacturing jobs were located in the 11 metropolitan areas, and they accounted for 89.4 percent of all jobs in the volatile durable goods industries. Of

---

9. Monroe County in southeast Michigan is also a metropolitan county but is part of the Toledo SMSA. It is not included in this study of Michigan metropolitan areas.

course, the Detroit SMSA stands out as the dominant area, accounting for 49.4 percent of total wage and salary employment in the state in 1977, 51.4 percent of manufacturing, 51.1 percent of nonmanufacturing, and 41.3 percent of government employment. This six-county area also accounted for 53.8 percent of all jobs in durable goods industries. The concentration of industrial activity in the metropolitan areas, therefore, suggests that the state's cyclically sensitive behavior emanates from swings in business activity within those areas. Because of its size, the Detroit SMSA tends to dominate overall activity in Michigan, but the behavior in other metropolitan areas contributes to the state's overall performance. Part I of this study deals with intrastate variations in short-run economic activity over a ten-year period.

TABLE 1

Michigan Standard Metropolitan Statistical Areas

Name	Counties included	1977 Population
Ann Arbor-Ypsilanti	Washtenaw	250,200
Battle Creek	Barry, Calhoun	182,000
Bay City	Bay	120,200
Detroit	Lapeer, Livingston, Macomb, Oakland, St. Clair, Wayne	4,370,200
Flint	Genesee, Shiawassee	514,400
Grand Rapids	Kent, Ottawa	575,600
Jackson	Jackson	149,900
Kalamazoo-Portage	Kalamazoo, Van Buren	268,100
Lansing-East Lansing	Clinton, Eaton, Ingham, Ionia	455,100
Muskegon-Norton Shores-Muskegon Heights	Muskegon, Oceana	179,000
Saginaw	Saginaw	226,700

Source: U.S. Department of Commerce, Bureau of the Census, *Current Population Reports,* P-26, No. 77-22.

PART I
METHODOLOGY AND COMPARATIVE ANALYSIS OF ECONOMIC INDICATORS

Measures of Local Business Activity and Intrastate Variations

Data deficiencies have been recognized for a long time by regional analysts as a serious problem that limits the scope of analysis at the subnational level. In examining short-run business conditions in states and local areas, one cannot draw upon the wide range of cyclical indicators that is available for the nation. For example, each month the U.S. Department of Commerce publishes over one-hundred cyclical indicators in *Business Conditions Digest*. These series are classified across seven major economic processes: (1) Employment and unemployment; (2) Production and income; (3) Consumption, trade, orders, and deliveries; (4) Fixed capital investment; (5) Inventories and inventory investment; (6) Price, costs, and profits; and (7) Money and credit. In addition, they are classified by their timing characteristics; that is, whether cyclical peaks and troughs in a specific series generally lead, lag, or are roughly coincident with those of aggregate business activity. This large set of cyclical indicators permits a fairly detailed analysis of economic behavior over various stages of the business cycle.

Such a comprehensive set of cyclical indicators is not available for subnational economies. In general, the number of reliable and fairly consistent time series varies inversely with the size of the region, so that fewer indicators are available at the local level compared to the number available for the state. However, even in cases where a good deal of local data exist, analysts must still grapple with changes resulting from modifications in data and/or area definitions. These are especially troublesome at the local level where such changes tend to occur with greater frequency and, in many cases, are far from trivial. The usual result is that intertemporal comparability is restricted, and this may have serious consequences for the type of time series analysis required to assess regional business conditions.

An example of an area definition change is the addition or subtraction of one or more counties to or from a metropolitan area, which effectively redefines the scope of the region's economy. A recent example of a data definition change is the shift in 1973 in measuring state and local unemployment rates from a place of work to a place of residence basis. This was done because the Comprehensive Employment and Training Act (CETA) passed in 1973 required that unemployment mean the same thing across states and local areas as well as for the nation as a whole. While this did not affect state data considerably, it significantly altered local unemployment data in areas subject to substantial in- and out-commuting. Unemployment data were revised back to 1970 but, because of the data definition change resulting from CETA, pre- and post-1970 jobless rates are not strictly comparable for local areas.[1] As a result, a longer historical perspective was lost.

Area and data definitions in this study were kept as consistent as possible, but in some cases, discrete breaks in local time series were unavoidable (see the discussion of banking below). The attempt to maintain consistency required a tradeoff in the length of some series. With few exceptions, time series used in this study began in 1970, thus limiting analysis of local behavior during the 1969-70 recession because cyclical peaks in specific series could not be determined. Nevertheless, consistent time series allowed measurements of local cyclical swings during the 1970s which were then used to compare intrastate performance.

Local Time Series

The titles, sources, and beginning year for each series used in this study are shown in Table 2. For each metropolitan area, 14 time series were assembled, covering local labor market, construction, and banking activity. Similar series were assembled for the State of Michigan as well, allowing comparisons between metropolitan performance and the state as a whole. Charts for each local time series appear in Part II.

Labor Market Data

Monthly data provided by the Michigan Employment Security

---

1. The Bureau of Labor Statistics of the U.S. Department of Labor was responsible for the changeover. The process of change is summarized by James R. Wetzel and Martin Ziegler, "Measuring Unemployment in States and Local Areas," *Monthly Labor Review*, June 1974, pp. 40-46.

Measures of Local Business Activity and Intrastate Variations 15

Commission (MESC) were used to assemble seven indicators of the employment/unemployment situation in each metropolitan area. Consistent monthly data based on the 1976 benchmark series of MESC were available from 1970 for five of the seven labor market series. Data on initial claims for unemployment insurance were available from 1968, as were average weekly hours of production workers in manufacturing industries for four metropolitan areas.

TABLE 2

Local Time Series and Data Sources

Series Title	Source	Initial Year
I. Labor Market		
Total Wage & Salary Employment	MESC	1970
Manufacturing Employment	MESC	1970
Nonmanufacturing Employment	MESC	1970
Government Employment	MESC	1970
Unemployment Rate	MESC	1970
Average Weekly Hours of Production Workers in Manufacturing[a]	MESC	1970
Initial Claims for Unemployment Insurance	MESC	1968
II. Construction		
New Building Permits, Private Housing	U.S. Dept. of Commerce, Bureau of the Census	1965
Construction Employment	MESC	1970
III. Banking[b]		
Demand Deposits	Federal Reserve Bank of Chicago	1970
Total Deposits	FED - Chicago	1970
Total Loans	FED - Chicago	1970
Commercial & Industrial Loans	FED - Chicago	1970
Consumer Installment Loans	FED - Chicago	1970

[a]For four SMSAs, a consistent average weekly hours series was available from 1968. This was also the case for the State of Michigan.

[b]Except for Battle Creek all series begin in 1970. Data for the Battle Creek SMSA are available from the fourth quarter of 1971.

The monthly data from MESC were converted into quarterly averages. The quarterly time series tended to be smoother and subject to somewhat smaller erratic movements than their monthly counterparts. These quarterly time series were sufficiently stable so that seasonal swings could be reliably estimated and removed. The seasonally adjusted quarterly employment and unemployment series were used to measure local cyclical swings.

Construction Data

Two series were used to measure local construction activity. Monthly construction employment data were provided by MESC and reflect the number of workers engaged in all types of building activity within a particular metropolitan area. Like the other employment series noted above, the monthly construction employment data were transformed into quarterly series and seasonally adjusted.

An Index of New Building Permits for private housing was developed for each area from monthly data published by the U.S. Department of Commerce *(Construction Reports—Housing Authorized by Building Permits and Public Contract, C-40)*. Deficiencies in the published data resulting from omission and definition changes were overcome to a great degree by supplementing the published information with reports from individual permit-issuing offices in the metropolitan areas. A major problem in constructing the new building permits time series was maintaining a consistent sample of governmental units over the 1965-77 period. In several cases this was possible. Ann Arbor-Ypsilanti, Battle Creek, and Kalamazoo-Portage have continuous series from 1965 and Saginaw from 1967. For the other metropolitan areas several governmental units were added in 1968, but this did not result in a sharp discontinuity because the additions were small relative to the total. Thereafter, the new building permits series are continuous. For each metropolitan area and the state, this series covers the last two national recession periods.

In almost all cases these monthly time series turned out to be highly erratic, exhibiting sizable changes from one month to the next. The irregular movements were reduced somewhat by averaging over quarters. However, while the quarterly series proved to be less erratic than the monthly series, they were still subject to sizable erratic swings and, consequently, estimates of seasonal movements turned out to be unreliable in many cases because of the highly unstable nature of the noncyclical movements in the series. Even in cases where seasonal estimates were fairly stable, the cyclical behavior was still obscured by irregular movements of considerable magnitude. To overcome these problems, a moving average was used to smooth the new building permits series and measure cyclical behavior. Of course, a disadvantage of this procedure is that the most recent data are lost, but this is not a major drawback for the historical analysis of this study.

To facilitate comparisons among the metropolitan areas, the computed moving average values were expressed in terms of their 1967 annual average. The charts in Part II show these quarterly index numbers (1967=100) as well as the unadjusted quarterly data (also in index form) for the 1965-77 period. At the time of this writing, only preliminary data for 1978 were available and, while they do not appear in the charts, they were used to measure the most recent behavior.

Banking Data

Commercial banking data for each metropolitan area were provided by the Federal Reserve Bank of Chicago. Five time series reflecting deposit and loan activity were used to measure local behavior. For all areas except Battle Creek data were available from 1970. The banking series for the Battle Creek SMSA begin in the fourth quarter of 1971.

Although the Federal Reserve series contain a great deal of information about commercial bank assets, liabilities, and types of deposits and loans, the data present two major problems for business conditions analysis. First, over the period examined, the number of banks included in the data collection procedure changed. Although this did not lead to a substantial break in the series in most areas, a discrete break was evident in several areas when banks were added. The most outstanding example of this procedure change occurred in the data for the Battle Creek SMSA between the first and the second quarters of 1973, when the number of reporting banks increased from three to six, resulting in a 60 percent jump in the current-dollar value of demand deposits in that area in one quarter. Such reporting changes were considerably less in other metropolitan areas but, overall, they imposed some upward bias on the banking data that could not be eliminated easily. Second, in some instances the number of quarterly observations was not large.

For example, only semiannual data were available from 1970 through 1972 for deposits and total loans. This restricted estimation of seasonal movements to the post-1972 years, a rather short time period. While the seasonal estimates for these series were stable, they must be regarded as preliminary and subject to revision as more data become available. Commercial and industrial loans and consumer installment loans were available only on a semiannual basis through 1975, quarterly thereafter. Consequently, the relatively short duration of those two series precluded any reliable estimation of seasonal factors, even of the most preliminary nature. Given these two limitations, the banking series behaved reasonably well over the period, with no extremely perverse movements evident.

The demand and total deposits series appearing in the charts in Part II, and on which measurements of cyclical behavior are based, were reduced to constant-dollar terms by deflating each quarterly value with the Consumer Price Index (CPI). The Detroit CPI was used to deflate the two Detroit deposits series, with the nationwide CPI being used for all other Michigan metropolitan areas. The deflated series were then placed in index form (1972=100) for comparative purposes. On the other hand, the loan series, which also appear in charts in Part II, were left in current-dollar terms.

Cyclical Behavior

The timing and amplitude of short-run fluctuations in the fourteen series for each metropolitan area were measured by determining cyclical peaks and troughs in specific series and then computing percent changes in series' values between those dates. Although seasonal swings were removed, as discussed above, the series were not adjusted for secular trend. Consequently, the amplitude measures reflect both trend and cycle behavior which represents actual experience, the primary concern of this study. Specific cycle turning points for these series were then compared with those of national business cycles so that, unless stated otherwise, classifications of leading, lagging, or roughly coincident local behavior refer to the timing of changes of movement of individual series for the metropolitan areas relative to cyclical swings in the national economy (reference cycles).

It should be noted that expansions and contractions in the local series generally conformed to national business cycle patterns. In

Measures of Local Business Activity and Intrastate Variations   19

other words, during a national recession those local series that would be expected to decline (manufacturing employment, for example) did so, while those that typically rise during a business slump (the unemployment rate, for example) behaved as expected in most cases. There were some notable exceptions, however. Two metropolitan areas, Ann Arbor-Ypsilanti and Flint, did not experience a cyclical decline in new building permits that conformed to the 1969-70 national recession. Also, the Ann Arbor-Ypsilanti, Muskegon-Norton Shores-Muskegon Heights, and Saginaw SMSAs experienced only a slowdown in growth, but not absolute decline, in the current-dollar volume of total loans during the 1973-75 recession. Finally, only 3 of 11 areas experienced a decline in government employment and only 7 of 11 areas had a falloff in the current-dollar volume of commercial and industrial loans during the 1973-75 recession that can be classified as cyclical.

While cyclical swings in business activity in Michigan metropolitan areas conformed to national patterns, there were substantial differences among the 11 areas in both the timing and amplitude of fluctuations in labor market, construction, and banking activity. These intrastate variations in performance are discussed below.

Timing

Table 3 shows the number of metropolitan areas where turning points in 14 specific indicators led, lagged, or were roughly coincident with the national business cycle peaks of the fourth quarter of 1969 and 1973. As discussed above, many of the local series began in 1970, so that timing comparisons with the business cycle peak of 1969 are rather limited. However, for those series that were available, average weekly initial claims for unemployment insurance, the average workweek of manufacturing workers (four areas only), and new building permits, leading behavior was the dominant characteristic. The longest leads occurred in the new building permits series, with the average lead time before the 1969 national peak amounting to 6.6 quarters for the 9 metropolitan areas experiencing a comparable cyclical decline. While that is a rather long average lead, it should be noted that the group average was influenced by very early downturns in the Battle Creek and Kalamazoo-Portage SMSAs. If those two areas are excluded, the average lead is 4.3 quarters for the other 7 metropolitan areas. That just about equals the four-quarter lead for the statewide series, and is only slightly above the three-quarter lead of the national series.

TABLE 3

**Leads and Lags at National Business Cycle Peaks
Number of SMSAs[a]**

Series	National Business Cycle Peak: 1969:4 Leading	Lagging	Roughly Coincident	Average lead (−) or lag (+) (quarters)	National Business Cycle Peak: 1973:4 Leading	Lagging	Roughly Coincident	Average lead (−) or lag (+) (quarters)
Total Wage & Salary Employment	NA	NA	NA	—	1	4	6	+0.8
Manufacturing Employment	NA	NA	NA	—	4	1	6	−0.5
Nonmanufacturing Employment	NA	NA	NA	—	1	7	2	+1.8
Government Employment	NA	NA	NA	—	1	0	2	−0.3
Unemployment Rate	NA	NA	NA	—	10	1	0	−1.4
Average Weekly Initial Claims for UI	11	0	0	−3.9	9	1	1	−1.7
Average Workweek, prod. workers, mfg.	4[b]	0	0	−3.3	10	0	0	−2.7
New Building Permits	9	0	0	−6.6	10	0	1	−4.0
Construction Employment	NA	NA	NA	—	8	2	1	−3.0
Deflated Demand Deposits	NA	NA	NA	—	11	0	0	−3.1
Deflated Total Deposits	NA	NA	NA	—	8	1	2	−1.7
Total Loans	NA	NA	NA	—	0	7	1	+1.9
Commercial & Industrial Loans	NA	NA	NA	—	2	4	1	+0.9
Consumer Installment Loans	NA	NA	NA	—	0	7	4	+1.6

Note: NA indicates that consistent local data were not available.

[a] Eleven SMSAs were considered for each series. In some cases, local peaks which conformed to national business cycle peaks were not discernible. Consequently, the total number of areas is less than eleven for some series.

[b] For four areas, Ann Arbor-Ypsilanti, Bay City, Jackson, and Saginaw, this series begins prior to 1969. For the other seven areas, a consistent average workweek series starts in 1970 so that peaks conforming to that national peak could not be determined.

Measures of Local Business Activity and Intrastate Variations   21

The same is true of the average lead of 3.9 quarters for average weekly initial claims in Michigan metropolitan areas, which only slightly exceeds the three-quarter lead of the national average weekly initial claims series. Although an average workweek series was available only for four metropolitan areas in the late 1960s, all four hit peaks before the recession began. From this limited set of data it appears that the timing characteristics for the group of metropolitan areas in Michigan were quite similar to their national counterparts on the average, although a good deal of variation exists.[2]

All 14 series were available for timing comparisons with the 1973 national business cycle peak. However, two series, government employment and commercial and industrial loans, showed little cyclical sensitivity among the metropolitan areas during the recession, with only three areas experiencing what could be classified as a cyclical contraction in government employment and only seven areas suffering an absolute decline in the current-dollar volume of commercial and industrial loans during the 1973-75 recession. No clear timing characteristics were evident for those two series.

Among the labor market indicators, the unemployment rate, average weekly initial claims for unemployment insurance, and the average workweek had cyclical turning points occurring in nearly all areas before the national recession began. Lead timing at business cycle peaks is a widely recognized characteristic of each of these series at the national level, so that the timing behavior in Michigan metropolitan areas at the 1973 peak was not out of line with expectations.

Peaks in total wage and salary employment were roughly coincident with the national business cycle peak in the majority of areas, but in four metropolitan areas, Grand Rapids, Kalamazoo-Portage, Lansing-East Lansing, and Muskegon-Norton Shores-Muskegon Heights, declines did not begin until the recession was well underway. Of the 11 metropolitan areas, only the Saginaw SMSA experienced a downturn in total wage and salary employment before the recession began, due to the early peak in local manufacturing employment, which accounted for just over

---

2. The lead for average weekly initial claims varied from seven quarters in the Ann Arbor-Ypsilanti SMSA to one quarter in the Jackson SMSA. The coefficient of variation for the 11 areas is .370. For the new building permits series the coefficient of variation is .785.

41 percent of wage and salary employment in that area in 1977. The other areas that experienced early peaks in manufacturing employment were Battle Creek, Jackson, and Kalamazoo-Portage, but the leads were not long, amounting to three, two, and one quarter, respectively. Except for the Saginaw SMSA, the metropolitan areas heavily dependent upon automobile production experienced downturns in total wage and salary and manufacturing employment that roughly coincided with the start of the national recession in 1973. In the nonmanufacturing sector, nine of the ten metropolitan areas experiencing a falloff in employment during the recession had peaks that occurred at or after the national business cycle peak. In other words, lagging behavior dominated the performance of nonmanufacturing employment.

In nearly all metropolitan areas in the state, the two indicators of local construction activity led the business cycle peak of 1973. As was the case with the 1969-70 recession, new building permits hit a peak well before the national business slump began. The average lead for the 11 metropolitan areas was four quarters, which equals the lead exhibited by the national series. Among the metropolitan areas in Michigan, timing ranged from leads of eight quarters in the Detroit and Flint SMSAs to coincidence in the Bay City SMSA. Employment in the construction industry also began falling in most areas before the recession started, for an average lead of three quarters. There were three notable exceptions to the leading behavior of construction employment: the Grand Rapids series, whose peak was roughly coincident, and the Kalamazoo-Portage and Muskegon-Norton Shores-Muskegon Heights SMSAs, where short lags were evident. Interestingly, those three metropolitan areas are located in western Michigan.

In the banking sector, deflated demand and total deposits led the national business cycle peak. Table 3 shows that the leading tendency was somewhat stronger for the former than the latter series, with all metropolitan areas in the state experiencing early peaks in deflated demand deposits. The average lead among the group was 3.1 quarters, with six areas leading by four quarters and five areas by two quarters. For the state as a whole, deflated demand deposits began declining four quarters before the national recession started. On the other hand, total loans and consumer installment loans, both in current-dollars, exhibited lags. Of the eight metropolitan areas where the volume of loans suffered absolute declines, seven did not begin falling until the recession was well underway. The longest lag, three quarters, occurred in the Bay

City, Grand Rapids, and Lansing-East Lansing SMSAs. The latter area, along with the Ann Arbor-Ypsilanti SMSA, also displayed the longest lag in consumer installment loans, four quarters. Overall, then, the peak timing behavior of the banking series in Michigan metropolitan areas was quite similar to the behavior of comparable national series.

Table 4 shows the timing characteristics of the same 14 local series at the national business cycle troughs of the fourth quarter of 1970 and the first quarter of 1975. Lagging and roughly coincident behavior dominated the trough patterns for total wage and salary and manufacturing employment. In 7 of the 11 metropolitan areas the upswing in manufacturing employment lagged behind the national business cycle recovery that began in early 1975. In five of those areas the lag was short—one quarter—but manufacturing employment lagged by five quarters in the Jackson SMSA and three quarters in the Muskegon SMSA. Combined with the two-quarter lead before the recession in the Jackson SMSA, the lag in manufacturing employment at the trough gives this area the longest cyclical downswing among the state's metropolitan areas, 12 quarters or 3 years. Furthermore, total wage and salary employment in the Jackson SMSA lagged considerably at the trough, resulting in the longest contraction for this series among the 11 metropolitan areas. The unemployment rate lagged in four areas at the 1970 trough, and in seven areas at the 1975 trough; behavior that is not particularly unusual, since the nationwide jobless rate also exhibited lags at both troughs. For the most part, the timing behavior of local labor market indicators reveals that turning points in these activities did not begin before the start of the two national business recoveries, and in many cases short lags were evident.

The outstanding timing characteristic in the construction sector is the lagging behavior of employment, particularly at the 1975 trough. In all 11 metropolitan areas construction employment lagged, with the average for the group amounting to 4.9 quarters. Lags of eight quarters occurred in the Ann Arbor-Ypsilanti, Jackson, Kalamazoo-Portage, and Lansing-East Lansing SMSAs, and for the state as a whole construction employment lagged by five quarters. Given the early peaks in most local areas, these long lags at the 1975 trough resulted in contractions in construction employment of fairly long duration in all Michigan metropolitan areas during the mid-1970s.

In the banking sector, the timing of troughs in the deflated

TABLE 4

Leads and Lags at National Business Cycle Troughs Number of SMSAs[a]

Series	National Business Cycle Trough: 1970:4			Average lead (−) or lag (+) (quarters)	National Business Cycle Trough: 1975:1			Average lead (−) or lag (+) (quarters)
	Leading	Lagging	Roughly Coincident		Leading	Lagging	Roughly Coincident	
Total Wage & Salary Employment	0	3	8	+0.5	0	5	6	+1.2
Manufacturing Employment	0	3	8	+0.5	0	7	4	+1.2
Nonmanufacturing Employment	2	3	6	+0.6	0	4	6	+0.4
Government Employment	1	3	1	+2.0	2	1	0	−0.3
Unemployment Rate	1	4	6	+0.8	0	7	4	+0.8
Average Weekly Initial Claims for UI	1	2	8	+0.1	2	2	7	+0.3
Average Workweek, prod. workers, mfg.	1	6	4	+1.3	3	3	4	−0.2
New Building Permits	8	1	0	−2.2	3	3	5	−0.3
Construction Employment	3	4	4	+0.3	0	11	0	+4.9
Deflated Demand Deposits[b]	8	0	2	−1.6	0	11	0	+5.9
Deflated Total Deposits[b]	10	0	0	−2.0	2	8	1	+2.0
Total Loans[b]	8	0	1	−1.8	1	7	0	+2.5
Commercial & Industrial Loans[b]	1	2	2	+0.4	2	5	0	+1.9
Consumer Installment Loans	4	1	4	−0.7	1	10	0	+1.7

[a] Eleven SMSAs were considered for each series. In some cases, local troughs which conformed to national business cycle peaks were not discernible. Consequently, the total number of areas is less than eleven for some series.

[b] In some cases, the initial value in the series was chosen as the specific cycle trough value. See charts in Part II.

deposit and total loans series differed considerably at the two national business cycle troughs. At the 1970 trough leading behavior dominated, with no lags evident. In contrast, lagging behavior characterized each of these series at the 1975 trough. The lags in deflated demand and total deposits, which were fairly long in some cases, appear to be unusual because of the lead tendency inherent in such series. However, it should be noted that the narrowly defined money supply, which includes cash plus demand deposits, also lagged nationally at that time and has recovered only slightly from its recession low. Therefore, the lags in the deflated demand deposits series in Michigan metropolitan areas are not unique. At least some of the lag is attributable to changes in financial practices and laws which have impacted on the use of demand deposits by individuals and businesses, resulting in a slowdown in the growth rate. But deflated total deposits, which include demand plus time deposits at commercial banks, also lagged at the 1975 trough. This appears to be the result of slow growth in the current-dollar volume of total deposits at commercial banks compared to the inflation rate.

The impression left by the timing performance of the banking indicators at the 1975 business cycle trough is that recovery in Michigan metropolitan areas began later than nationwide. The lags in some of the labor market and construction indicators support that conclusion. Thus, the duration of the slump in business activity in Michigan metropolitan areas appears to have been longer than the national recession. During the period examined, the earliest signs of local business downturns were evident in the indicators of construction activity in most metropolitan areas. But several indicators of local labor market and banking activity also exhibited early peaks.

Amplitude

The magnitudes of the cyclical swings in the 14 local economic indicators are given in Tables 5, 6, and 7. Table 5 presents the amplitudes of the cyclical expansion phases that generally conformed to the national business cycle expansion of 1970-73. The measures shown in the table represent absolute values of the percent change in each series from its initial trough to terminal peak. Table 6 shows the absolute values of percent changes (peak to trough) in each series for the cyclical contraction conforming to the 1973-75 national recession. Finally, Table 7 gives the absolute values of percent changes in each series for the expansion phase conforming to the most recent national business cycle expansion

TABLE 5

Amplitude of Expansions in Specific Series Conforming to the National Business Expansion of 1970-73[a]

Series	Ann Arbor Ypsi.	Battle Creek	Bay City	Detroit	Flint	Grand Rapids	Jackson	Kalamazoo Port.	Lans. East Lans.	Musk.-N.S. Musk. Hts.	Saginaw	Mich.
Total Wage & Salary Employment	20.4	9.1	19.8	13.4	40.7	19.4	17.4	17.1	23.9	13.5	31.3	15.2
Mfg. Employment	38.0	8.9	20.9	19.4	93.5	24.0	21.7	14.0	57.4	14.2	64.7	23.2
Nonmfg. Employment	34.5	9.7	23.0	12.0	11.4	18.1	23.2	b	22.6	14.2	19.8	14.5
Govt. Employment	b	14.9	b	b	21.5	b	b	b	b	b	b	b
Unemployment Rate	41.7	29.7	48.2	35.8	57.5	34.2	46.7	38.4	45.0	43.0	37.3	35.7
Average Weekly Initial Claims for UI	55.8	40.1	64.3	60.6	69.7	69.5	66.5	44.4	64.6	58.1	70.4	52.6
Average Workweek, prod. workers, mfg.	16.0	9.3	b	15.4	18.8	3.8	14.9	4.4	19.3	12.1	18.8	11.3
New Building Permits	130.1	259.0	175.1	60.9	104.5	107.4	135.2	200.0	148.4	115.8	60.0	61.3
Construction Employm't	30.0	23.5	70.0	9.6	21.6	32.1	33.3	29.4	21.6	43.8	32.1	15.2
Deflated Demand Deposits[c]	13.0	d	15.5	26.3	24.5	21.2	20.2	40.2	26.7	19.8	23.4	24.2[e]
Deflated Total Deposits[f]	19.3	d	22.8	21.6	33.5	26.5	26.1	33.5	17.3	34.1	28.2	20.0[e]
Total Loans[f]	b	d	26.2	47.2	62.5	59.6	56.6	69.6	40.6	b	b	b
Comm. & Ind. Loans[e]	b	d	16.8[f]	40.2[f]	44.4[f]	86.0	50.5	b	b	b	91.1	39.7
Consumer Installment Loans[e]	48.0	d	47.3[f]	68.8	58.0	77.6	75.4[f]	b	72.9[f]	86.3	142.7[f]	59.9

[a] Amplitude measures represent percent changes from trough to peak, disregarding the sign, for specific phases conforming to the 1970-73 national business cycle expansion.
[b] Expansion phase not discernible.
[c] Except for Ann Arbor-Ypsilanti and Saginaw, the beginning value of the series was chosen as the initial trough value for the expansion period.
[d] For Battle Creek, data prior to 1971 were not available so that the entire expansion phase could not be measured.
[e] Based on seasonally unadjusted data.
[f] Beginning value of the series chosen as initial trough value for the expansion period.

TABLE 6

Amplitude of Contractions in Specific Series Conforming to the 1973-75 National Recession[a]

Series	Ann Arbor Ypsi.	Battle Creek	Bay City	Detroit	Flint	Grand Rapids	Jack-son	Kala-mazoo Port.	Lans. East Lans.	Musk.-N.S.-Musk. Hts.	Sagi-naw	Mich.
Total Wage & Salary Employment	8.0	6.3	6.0	8.1	13.4	5.1	8.3	3.2	3.9	4.3	7.7	6.9
Mfg. Employment	26.0	15.7	19.2	20.1	25.5	18.3	25.5	13.6	24.5	15.3	20.6	18.7
Nonmfg. Employment	5.1	2.2	2.2	3.6	5.5	1.2	4.6	b	1.7	2.4	3.7	3.4
Govt. Employment	b	8.6	b	b	8.3	b	4.7	b	b	b	b	b
Unemployment Rate	231.0	146.2	150.9	140.4	220.4	148.0	212.5	160.0	204.5	153.8	192.9	148.1
Average Weekly Initial Claims for UI[c]	638.8 (337.2)	396.7 (104.6)	301.7 (324.8)	375.1 (237.3)	522.7 (478.4)	452.6 (137.7)	282.0 (253.8)	258.0 (176.9)	222.9 (342.6)	198.3 (227.2)	503.9 (456.1)	301.1 (193.2)
Average Workweek, prod. workers, mfg.	13.6	5.3	b	12.5	14.5	5.3	7.6	7.0	14.6	7.1	12.5	9.7
New Building Permits[c]	88.6 (b)	84.5 (66.6)	72.3 (54.4)	67.5 (29.3)	79.6 (b)	46.3 (43.0)	33.9 (33.3)	75.3 (58.3)	71.2 (63.4)	72.7 (33.1)	73.2 (31.5)	56.2 (21.5)
Construction Employm't	38.4	33.3	47.1	24.9	33.9	20.7	45.0	18.2	24.2	13.0	40.5	23.5
Deflated Demand Deposits	21.6	28.9	31.8	29.0	26.7	24.4	26.4	25.1	28.6	26.4	19.9	31.1[d]
Deflated Total Deposits	8.4	15.7	17.8	11.8	13.4	9.1	10.3	14.7	15.7	11.0	6.0	11.2[d]
Total Loans	b	9.8	4.1	7.0	4.8	4.8	6.4	5.1	4.8	b	b	b
Comm. & Ind. Loans[d]	b	26.1	21.4	7.3	6.3	5.6	18.3	b	b	b	12.2	0.8
Consumer Installment Loans[d]	4.3	32.5	9.0	9.8	6.0	8.7	14.4	b	8.4	6.3	6.2	6.6

[a] Amplitude measures represent percent changes from peak to trough, disregarding the sign, for specific phases conforming to the 1973-75 national business cycle contraction.
[b] Contraction phase not evident.
[c] Numbers in parentheses are comparable measures for the 1969-70 recession.
[d] Measures based on seasonally unadjusted data.

27

TABLE 7
Amplitude of Expansions in Specific Series Conforming to the National Business Expansion Starting in 1975[a]

Series	Ann Arbor Ypsi.	Battle Creek	Bay City	Detroit	Flint	Grand Rapids	Jackson	Kalamazoo Port.	Lans. East Lans.	Musk. N.S.- Musk. Hts.	Saginaw	Mich.
Total Wage & Salary Employment	21.4	10.5	19.5	13.5	27.9	17.5	8.8	12.9	17.2	10.8	23.3	15.5
Mfg. Employment	46.1	4.9	39.3	18.6	40.5	29.0	17.8	18.1	35.3	11.7	34.5	20.0
Nonmfg. Employment	15.8	15.5	15.2	13.2	25.4	12.9	10.9	b	13.9	15.6	14.7	14.9
Govt. Employment	b	20.8	b	b	24.5	b	9.9	b	b	b	b	b
Unemployment Rate	64.0	50.8	57.3	48.8	59.5	61.3	55.2	53.0	55.2	52.1	58.5	50.0
Average Weekly Initial Claims for UI	77.3	66.5	72.8	62.2	75.9	67.4	59.9	59.8	51.8	48.4	85.6	59.2
Average Workweek, prod. workers, mfg.	13.2	3.2	b	11.8	22.7	7.4	10.8	6.0	14.6	8.4	19.9	9.3
New Building Permits	162.9	240.3	143.9	93.8	194.7	82.9	141.8	212.1	132.1	196.5	79.9	60.1
Construction Employm't	37.5	21.4	100.0	52.0	87.8	29.5	54.5	19.4	14.9	50.0	40.9	47.5
Deflated Demand Deposits	14.8	19.0	14.3	16.7	26.7	22.2	15.4	11.2	9.5	6.9	21.6	19.4[c]
Deflated Total Deposits	19.6	9.7	15.2	8.5	18.8	14.8	6.9	9.1	7.1	8.1	17.1	9.5[c]
Total Loans	b	41.7	89.0	29.2	42.5	40.7	27.3	32.1	26.3	b	b	b
Comm. & Ind. Loans[c]	b	53.9	118.1	32.2	b	57.7	b	b	b	b	87.9	33.4
Consumer Installment Loans[c]	72.1	49.7	101.0	54.7	80.9	49.6	35.5	b	34.8	40.0	54.3	51.4

[a] Amplitude measures represent percent changes from trough to the most recent high or low, disregarding the sign, for specific phases conforming to the national business cycle expansion that began in early 1975. These measures are preliminary since expansion phases may be incomplete.
[b] Expansion phase not discernible.
[c] Based on seasonally unadjusted data.

Measures of Local Business Activity and Intrastate Variations   29

that began in the first quarter of 1975. The measures in Table 7 represent changes from the trough value to a most recent high or low, depending upon whether the series is a positive or inverse series. In many cases, these measures may reflect incomplete phases, because it is not yet clear whether the national and local expansions have ended and business contractions are underway. The measures presented in the three tables reveal wide disparities in performance among Michigan metropolitan areas during phases of the business cycle.

Note that among the labor market indicators, manufacturing employment and the average workweek of manufacturing workers exhibited sizable upswings and downswings in the Ann Arbor-Ypsilanti, Flint, Lansing-East Lansing, and Saginaw SMSAs. This is not at all surprising, considering that each of those areas is heavily dependent on the automobile industry. During the 1973-75 recession, manufacturing employment fell 26 percent in Ann Arbor-Ypsilanti, 25.5 percent in Flint (and Jackson), 24.5 percent in Lansing-East Lansing, and 20.6 percent in Saginaw, while hours worked fell 13.6, 14.5, 14.6, and 12.5 percent, respectively. Table 5 and 7 show that the two expansions were also relatively strong in each of those four areas, resulting in a high degree of overall cyclical sensitivity in their manufacturing sectors. In contrast, two of the areas least dependent on durable goods production, the Battle Creek and Kalamazoo-Portage SMSAs, exhibited the lowest degree of cyclical sensitivity within their manufacturing sectors. During the 1973-75 recession manufacturing employment fell 13.6 percent in the Kalamazoo-Portage SMSA, the mildest contraction among Michigan metropolitan areas and nearly one-half the magnitude of the severe slump that occurred in the Ann Arbor-Ypsilanti SMSA.

The swings in total wage and salary employment do reflect the varying degrees of volatility of local manufacturing employment, but not perfectly. Total wage and salary employment in the Flint SMSA exhibited the most cyclical sensitivity, total wage and employment in the Kalamazoo-Portage SMSA the least. Despite having highly volatile manufacturing sectors, contractions in total wage and salary employment in the Ann Arbor-Ypsilanti and Lansing-East Lansing SMSAs were relatively mild due to the moderating influence of the large and fairly stable nonmanufacturing and government sectors in each area.

The Ann Arbor-Ypsilanti, Flint, and Saginaw SMSAs also exhibited sizable swings in average weekly initial claims for unemployment insurance. Data from 1968 were available so that behavior during the two recession periods can be compared. During the 1969-70 recession average weekly initial claims rose 337.2 percent in the Ann Arbor-Ypsilanti SMSA, 478.4 percent in the Flint SMSA, and 456.1 percent in the Saginaw SMSA, with the areas ranking fourth, first, and second, respectively. The Lansing-East Lansing SMSA ranked third in this respect, as average weekly initial claims rose 342.6 percent. Table 6 shows that the Ann Arbor-Ypsilanti, Flint, and Saginaw SMSAs ranked first, second, and third, respectively, in terms of the cyclical rise in average weekly initial claims during the 1973-75 recession. Although that business slump was more severe than the 1969-70 episode, average weekly initial claims actually rose less in the Bay City, Lansing-East Lansing, and Muskegon SMSAs. In all other areas, the state, and the nation, the cyclical upswing in average weekly initial claims during the 1973-75 recession was substantially larger than that of the 1969-70 period (see charts in Part II).

During the two business expansion periods, sizable declines in average weekly initial claims accompanied employment upswings in each metropolitan area. Tables 5 and 7 show that so far during the recent expansion (1975-78) the falloff in average weekly initial claims exceeds that of the previous expansion (1970-73) in all areas except Grand Rapids, Jackson, Lansing-East Lansing, and Muskegon. Despite declines since the last recession, however, average weekly initial claims in all Michigan metropolitan areas at the end of 1978 remained above their pre-recession lows. The business expansions in these local areas, therefore, have failed to pull average weekly initial claims down enough to offset the sizable runups that occurred during the recession. Consequently, this series exhibits an upward trend over the entire period in all Michigan metropolitan areas and the state as a whole, a phenomenon also noticeable nationwide.

The behavior of the local unemployment rates is similar in that the recession led to sizable upswings, and, in most cases, the declines during the 1975-78 expansion have not offset the previous increases. The hardest hit areas during the recession, i.e., those with large peak to trough percent changes, tended to be the areas where the slump in manufacturing employment was particularly severe: Ann Arbor-Ypsilanti, Flint, Jackson, Lansing-East Lansing, and Saginaw. In fact, the correlation between the amplitudes of those two series for the recession period is +.85, which indicates a strong association between the magnitudes of the contraction phases of manufacturing employment and the jobless rate across Michigan metropolitan areas. This positive relationship is also evident during expansions, although the degree of association is not as strong as that of the recession period. What this relationship suggests is that metropolitan areas in Michigan which have a large and highly volatile manufacturing base are also susceptible to sizable swings in unemployment. The Ann Arbor-Ypsilanti and Flint SMSAs are excellent examples of this type of local instability. Both areas experienced sharp declines in manufacturing employment during the 1973-75 recession resulting from the falloff in activity in the automobile industry. The unemployment rate in the Ann Arbor-Ypsilanti SMSA rose from 4.2 percent to 13.9 percent, and in the Flint SMSA it jumped from 5.4 percent to 17.3 percent. The latter represents the highest quarterly jobless rate among the Michigan metropolitan areas during the recession. At the other extreme is the Kalamazoo-Portage SMSA, which experienced the mildest slump in manufacturing employment and whose jobless rate rose from 4.5 percent to 11.7 percent. Although the latter was well above the national average, it was below the statewide high of 13.4 percent and was the lowest among the 11 metropolitan areas. By the end of 1978 the unemployment rate in nearly all Michigan metropolitan areas had not fallen below pre-recession lows. The only exception is the Grand Rapids SMSA, where the jobless rate hit a seasonally adjusted 4.8 percent in the first quarter of 1978, just below its pre-recession low of 5.0 percent.

The two indicators of local construction activity reveal a high degree of cyclical sensitivity in Michigan metropolitan areas. In all areas the total amplitude of the fluctuations in the Index of New Building Permits exceeded that of the state and the nation. With few exceptions this was also the case for specific contraction and expansion phases. The two recession periods are quite revealing. All metropolitan areas except Ann Arbor-Ypsilanti and Flint experienced larger downswings in new building permits than the state as a whole during the 1969-70 recession. The Ann Arbor-Ypsilanti and Flint SMSAs did experience contractions from the mid- to the late 1960s, but they did not conform to the national recession (see charts in Part II). As Table 6 shows, only the Grand Rapids and Jackson SMSAs suffered less severe declines than the state during the 1973-75 slump. The most severe contractions occurred in the Ann Arbor-Ypsilanti, Battle Creek, and Flint SMSAs although, as Table 6 shows, many other metropolitan areas experienced declines that were only slightly milder than the state as a whole. With only two exceptions, then, the drop in the Index of New Building Permits was sharp in Michigan metropolitan areas during the mid-1970s. Table 7 shows, however, that the upswings following the last recession were also sizable, with the largest relative increases occurring in the Battle Creek and Kalamazoo-Portage SMSAs. Nevertheless, all Michigan metropolitan areas except Jackson had index levels in 1978 below their pre-recession peaks. This situation is not unusual, since both the statewide and nationwide indexes have also failed to surpass their previous peaks during the recent business expansion.

The behavior of construction employment was similar to but not exactly the same as new building permits. The magnitudes of the upswings in this local series during the 1970-73 business expansion exceeded the statewide gain in all metropolitan areas except Detroit (see Table 5). Table 6 shows that the slump during the mid-1970s was more severe than in the state as a whole in all metropolitan areas except Grand Rapids, Kalamazoo-Portage, and Muskegon. Note that the Grand Rapids SMSA was the only metropolitan area in the state to experience a relatively mild slump in construction employment and new building permits. Interestingly, the Jackson SMSA suffered the largest relative drop in construction employment and the mildest falloff in new building permits among the 11 metropolitan areas. Table 7 shows that in only 5 of 11 metropolitan areas has construction employment expanded more than it has statewide from the trough to the most recent high. In 9 of 11 areas the most recent expansion of construction employment exceeds the 1970-73 upswing. The Grand Rapids and Kalamazoo-Portage SMSAs are the two exceptions, but those areas also experienced relatively mild contractions during the recession. By the end of 1978, construction employment had failed to reach pre-recession levels in four areas: Ann Arbor-Ypsilanti, Battle Creek, Jackson, and Lansing-East Lansing, and the latter area has exhibited the most sluggish upswing (see Part II).

In the banking sector, deflated demand and total deposits suffered the most severe downswing during the recession in the Bay City SMSA, with declines of 31.8 percent and 17.8 percent, respectively. The Saginaw SMSA, on the other hand, experienced the mildest setbacks in the two deposits series, despite the rather sizable falloff in employment that occurred in that area. The Flint SMSA, another volatile area in terms of employment and building activity, did not exhibit extreme declines in banking activity. Deflated total deposits fell 13.4 percent in the Flint SMSA, and the current dollar volume of total loans fell 4.8 percent. In comparison, the Kalamazoo-Portage SMSA, a much more stable area in terms of employment, experienced slightly greater declines in deflated total deposits and total loans during the recession, 14.7 percent and 5.1 percent, respectively. The Battle Creek SMSA, which also experienced relatively mild declines in employment, suffered substantial drops in total loans, commercial and industrial loans, and consumer installment loans. In the case of consumer installment loans, the Battle Creek SMSA experienced a decline of 32.5 percent in current-dollar volume, the largest contraction by far among the 11 metropolitan areas, whereas the adjacent Kalamazoo-Portage SMSA suffered no decline during the recession. Also, along with several other areas, the Kalamazoo-Portage SMSA experienced no measurable cyclical decline in commercial and industrial loans.

Table 7 shows that banking activity picked up substantially during the recent business expansion, except for deflated demand deposits. The latter have increased at a rather sluggish pace, and by mid-1978 they were still below pre-recession peaks in all Michigan metropolitan areas. While deflated total deposits had risen during the business expansion, they were below their previous peaks in all areas except the Ann Arbor-Ypsilanti, Grand Rapids, and Saginaw SMSAs. The current-dollar volume of loans has expanded considerably in all areas since the last recession. Table 7 shows that the greatest relative increase occurred in the Bay City SMSA, where the current-dollar volume of total loans has risen 89 percent since the recession, and commercial and industrial loans and consumer installment loans have more than doubled.

In summary, then, banking activity in Michigan metropolitan areas exhibited varying degrees of cyclical sensitivity during the period examined. Only three areas, Ann Arbor-Ypsilanti, Muskegon, and Saginaw, escaped absolute declines in the current-dollar volume of total loans during the 1973-75 recession. Also, the Kalamazoo-Portage SMSA was the only metropolitan area where consumer installment loans did not fall during that slump. All areas, however, experienced declines in the deflated values of total and demand deposits, with the latter exhibiting sluggish recoveries.

The measurements presented above reveal differences in business conditions among Michigan metropolitan areas from the late 1960s to 1978. While a wide range of economic indicators was assembled and their behavior examined, the number of years covered in this study represents a rather short time period that encompassed one national business cycle peak in late 1973 and two troughs, one occurring in late 1970 and another in early 1975. Local behavior during the two business expansion periods and one recession period was analyzed, and the results reveal varying degrees of cyclical sensitivity among Michigan metropolitan areas in terms of labor market, construction, and banking activity. It should be noted that the findings of this study refer only to local business conditions during those years, and caution is urged regarding generalizations or forecasts about the local economic activity in the future from the historical analysis above. Certainly, this study provides a good deal of information about the metropolitan economies in Michigan and develops a solid methodological base from which to examine and compare current business conditions within the state. As indicated previously, use of the economic indicators developed in this study for current business conditions analysis requires updating the data in light of recent revisions by reporting agencies and then reestimating the seasonal movements so that the indexes are derived not only from observations covering a longer time span but also reflect the most recent conditions to the extent that this is possible. Both of those changes are likely to improve the reliability of such estimates, which are crucial to the assessment of current conditions and forecasts of the direction of local economies.

PART II
MICHIGAN
METROPOLITAN AREAS

Ann Arbor-Ypsilanti SMSA

The Ann Arbor-Ypsilanti SMSA is the sixth largest among Michigan's 11 metropolitan areas in terms of population. This metropolitan area, which encompasses Washtenaw County, experienced a population increase of 6.9 percent from 1970 to 1977, as its population rose from 234,103 in 1970 to 250,200 in 1977. The population increase in this area slightly exceeded the nationwide growth of 6.4 percent and was considerably above the 2.8 percent increase registered in Michigan over that period. Among the state's 11 metropolitan areas, the population growth in the Ann Arbor-Ypsilanti SMSA was exceeded only by the 7.3 percent increase that occurred in the Lansing-East Lansing SMSA.

The area also experienced relatively strong growth of personal income. From 1969 to 1976, its personal income grew at an annual rate of 9.8 percent, highest among the state's 11 metropolitan areas. Moreover, income growth in the Ann Arbor-Ypsilanti SMSA exceeded the nationwide rate of 9.3 percent and the average rate for all U.S. SMSA's of 9.0 percent. In 1976, total personal income in this metropolitan area was $1,790 million. Four metropolitan areas in Michigan exceeded that total: Detroit, Flint, Grand Rapids, and Lansing-East Lansing.

Per capita income in the area is relatively high. In 1976 it was estimated at $7,215 compared to $6,396 nationwide, $6,757 in Michigan, and an average of $6,824 for all U.S. SMSAs. Among Michigan metropolitan areas, only the Detroit SMSA, with a per capita income level of $7,496, ranked ahead of the Ann Arbor-Ypsilanti SMSA. The relatively high income level in this metropolitan area was the result of above-average level of wages paid by the durable goods industries, including automobile-related manufacturing firms located there, combined with the salaries of professional, managerial, and technical workers employed in Washtenaw County.

The Ann Arbor-Ypsilanti SMSA has a fairly diversified economy, with total wage and salary employment about equally distributed with one-third in manufacturing, one-third in nonmanufacturing, and one-third in government employment. Table A1 shows that in 1972, 34.1 percent of the area's wage and salary employment was accounted for by manufacturing. By 1977, that share had fallen to 32.4 percent, exactly the same as the proportion statewide. Durable goods production, primarily linked to the automobile industry, dominates the area's manufacturing sector. Industries producing durable goods accounted for 30.2 percent and 29.2 percent of total wage and salary employment in the area in 1972 and 1977, respectively. As Table A1 shows, those shares were slightly above statewide proportions.

TABLE A1

Percentage Distribution of Total Wage and Salary Employment
Ann Arbor-Ypsilanti SMSA and Michigan, 1972 and 1977

Item	Ann Arbor-Ypsilanti SMSA 1972	Ann Arbor-Ypsilanti SMSA 1977	Michigan 1972	Michigan 1977
Total Wage and Salary....	100.0%	100.0%	100.0%	100.0%
Manufacturing	34.1	32.4	35.1	32.4
Durables	30.2	29.2	28.1	25.9
Nondurables	3.9	3.2	7.0	6.5
Nonmanufacturing	30.8	35.0	48.0	49.6
Government	35.1	32.6	16.9	18.0

Source: Michigan Employment Security Commission.

Government employment accounted for 35 percent of total wage and salary employment in the area in 1972, and 32.6 percent in 1977. Despite a downward shift in the share of jobs, Table A1 shows that a much higher proportion of workers in this metropolitan area are employed by the public sector than is the case for Michigan as a whole. The size of the government sector in the Ann Arbor-Ypsilanti SMSA is attributable to the fact that two state universities are located there, the University of Michigan in Ann

Arbor and Eastern Michigan University in Ypsilanti. Both the size and stability of this area's public sector temper the more unstable situation in local manufacturing industries resulting from a heavy concentration in automobile-related production.

Overall, the major structural differences in employment between the state and this metropolitan area are the latter's slightly greater concentration of workers in durable goods manufacturing and the significantly larger proportion of its wage and salary jobs in government. In fact, only the Lansing-East Lansing SMSA, site of the state capitol, has a larger share of employment accounted for by the public sector.

Business Conditions in the Ann-Arbor-Ypsilanti SMSA

Labor Market Conditions

Total wage and salary employment and its major components— manufacturing, nonmanufacturing, and government employment— are shown by quarter in Chart A1. For the 1970-78 period, the overall behavior of total wage and salary and manufacturing employment conformed to national business cycles. In other words, these two key employment series increased during national expansion periods and declined during recessions. As Chart A1 shows, the peaks and troughs in the two series were roughly coincident with turning points in aggregate business activity in the nation. In addition, while nonmanufacturing employment displayed some cyclical sensitivity, government employment did not.

During the expansion period from the fourth quarter of 1970 to the fourth quarter of 1973, total wage and salary employment rose 20.4 percent in the metropolitan area. That increase resulted from a rise of 38.0 percent and 30.2 percent in manufacturing and nonmanufacturing employment, respectively, over that same period. Interestingly, government employment in the Ann Arbor-Ypsilanti SMSA remained relatively flat in the period between the two recessions.

As previously noted, manufacturing in this metropolitan area is heavily weighted toward durable goods production in those industries most seriously affected during economic slowdowns. Consequently, manufacturing employment in the area experienced a severe contraction during the 1973-75 recession, falling 26 percent from the fourth quarter of 1973 to the first quarter of 1975. That was the largest relative decline in manufacturing employment

Ann Arbor-Ypsilanti SMSA

CHART A 1
WAGE AND SALARY EMPLOYMENT, ANN ARBOR-YPSILANTI SMSA
(Seasonally adjusted)

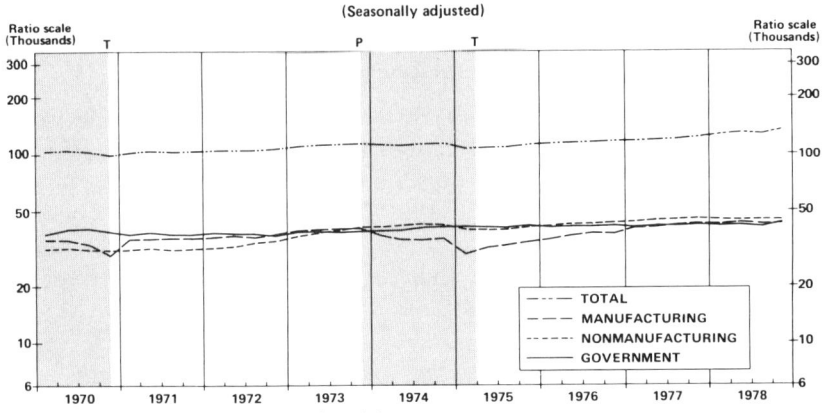

Source: Michigan Employment Security Commission.
Notes: Seasonal adjustment by the W.E. Upjohn Institute.
Shaded areas indicate national recession periods as defined by the National Bureau of Economic Research, Inc.
P = peak and T = trough.

CHART A 2
UNEMPLOYMENT RATE
ANN ARBOR - YPSILANTI SMSA
(Seasonally adjusted)

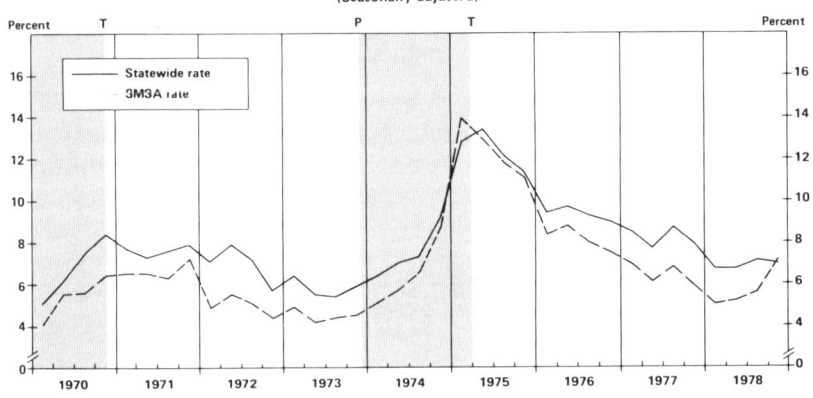

Source: Michigan Employment Security Commission.
Note: Seasonal adjustment by the W.E. Upjohn Institute.
Shaded areas indicate national recession periods as defined by the National Bureau of Economic Research, Inc.
P = peak and T = trough.

among the state's 11 metropolitan areas during this period. The contraction in total wage and salary employment was considerably less, however, 8 percent from peak to trough. The less severe slump in total wage and salary employment was due to the fact that the sizable decline in manufacturing employment was tempered by a brief and relatively mild contraction in the nonmanufacturing sector and an actual rise in government employment. And of major importance in this metropolitan area was the nearly equal distribution of total wage and salary employment among the three major sectors. Because of this, the behavior of the area's highly cyclically sensitive manufacturing sector has less of an impact on aggregate employment.

By the end of 1978, the employment series shown in Chart A1 were all above their pre-recession levels. Over nearly a four-year period beginning in early 1975, total wage and salary employment rose 21.4 percent, with manufacturing employment up 46.1 percent, nonmanufacturing employment, 15.3 percent, and government employment, 8.9 percent. Except for the nonmanufacturing sector, employment increased more during the 1975-78 period than during the previous expansion period of 1970-73.

Growth rates for the entire 1970-78 period are given in Table A2. Total employment growth in the Ann Arbor-Ypsilanti SMSA exceeded the statewide rate, despite the fact that government employment grew at less than one-half the Michigan rate over the period. Manufacturing and nonmanufacturing employment growth exceeded the statewide performance by considerable margins. Moreover, the Ann Arbor-Ypsilanti SMSA registered the highest growth rates for manufacturing and nonmanufacturing employment among the 11 metropolitan areas in Michigan. For the 1970-78 period, the 2.8 percent average annual growth of total wage and salary employment was exceeded only by the 3.1 percent growth in the Grand Rapids SMSA.

The seasonally-adjusted unemployment rate for the Ann Arbor-Ypsilanti SMSA is shown in Chart A2 along with the jobless rate for Michigan as a whole. For the most part, the local jobless rate remained below that of the state during the 1970-78 period. However, during the 1973-75 recession, the area's unemployment rate rose from a low of 4.2 percent in the second quarter of 1973 to a high of 13.9 percent in the first quarter of 1975, with the latter exceeding the statewide average. It is interesting to note that the percentage increase in this upswing in the unemployment rate in the Ann Arbor-Ypsilanti SMSA was exceeded only by that of the Flint

42 Ann Arbor-Ypsilanti SMSA

SMSA (from a low of 5.4 percent in the third quarter of 1973 to a high of 17.3 percent in the first half of 1975). Two other SMSAs had seasonally-adjusted unemployment rates higher than that of the Ann Arbor-Ypsilanti SMSA at the depth of the 1973-75 downturn. Bay City at 14.3 percent and Muskegon at 16.5 had both recorded greater levels of joblessness.

TABLE A2

Average Annual Growth Rates of Selected Labor Market Indicators Ann Arbor-Ypsilanti SMSA and Michigan[a]

(percent)

Indicator	Ann Arbor-Ypsilanti SMSA	Michigan
Total wage and salary employment	2.8	1.8
Manufacturing employment	2.1	0.2
Nonmanufacturing employment	5.2	2.6
Government employment	1.4	3.0
Civilian labor force	3.3	2.0
Unemployment rate	3.7	2.8
Average weekly initial claims for UI[b] . .	11.1	7.3
Average workweek, production workers, mfg.[b]	0.4	0.2

[a]Except where indicated otherwise, estimated growth rates are based on log-linear trends for the 1970-78 period.

[b]Computed for the 1968-78 period.

Another notable characteristic of the area's unemployment rate is its secular uptrend over the period. On an annual basis, the jobless rate grew at a 3.7 percent rate, which is higher than the state growth rate of 2.8 percent. Nevertheless, the area's uptrend was less than the 4.3 percent growth rate for the nation over the 1970-78 period. To some degree, this growth of the unemployment rate in the area was attributable to increases in the labor force. As Table A2 shows, the local labor force grew at an annual rate of 3.3 percent, compared to a more modest statewide rate of 2.0 percent.

Another key unemployment indicator, average weekly claims for unemployment insurance, is shown in Chart A3. For comparative purposes, statewide claims are also presented. With only minor differences, the behavior of the local and statewide series was very similar. Both began their cyclical upswings several quarters before the national recessions started. This leading tendency at peaks is also a characteristic of the national series of average weekly initial claims for unemployment insurance covering all state programs. In the Ann Arbor-Ypsilanti SMSA, initial claims began falling off at about the time that the business expansion started.

Despite general similarities in behavior, however, initial claims for unemployment insurance were more cyclically sensitive in this local area than for the state as a whole. For example, during the 1969-70 recession, average weekly initial claims more than tripled in the Ann Arbor-Ypsilanti SMSA compared to a doubling for the state. During the more severe recession in 1973-75, initial claims in the local area rose to a level more than six times higher than their low in the first quarter of 1973. That sizable increase was the largest among Michigan metropolitan areas, exceeding by a considerable margin the statewide tripling of claims. As Chart A3 shows, initial claims have fallen throughout most of the recent expansion period, but they have not moved below pre-recession levels. For the entire 1968-78 period, average weekly initial claims in the Ann Arbor-Ypsilanti SMSA grew at an annual rate of 11.1 percent, the highest growth rate among the 11 metropolitan areas and, as Table A2 shows, considerably above the growth for Michigan as a whole.

As indicated above, manufacturing employment in the Ann Arbor-Ypsilanti SMSA has been quite volatile. Therefore, it is not surprising that initial claims for unemployment insurance also exhibited a high degree of cyclical sensitivity, because nearly all manufacturing workers were covered under the state program. A further manifestation of cyclical sensitivity in this area's manufacturing sector is found in the behavior of the average workweek of production workers. Chart A4 shows that this series conformed to national business cycle patterns over the 1968-78 period but with a lead, particularly at peaks. Since turning points in the area's total wage and salary and manufacturing employment were roughly coincident with peaks and troughs in national business activity, changes in the cyclical phase of the average workweek tended to foreshadow those of the two employment series. Therefore, the average workweek, like average weekly initial claims for unemployment insurance, appears to be a leading indicator in the Ann Arbor-Ypsilanti SMSA. The magnitude of the cyclical swings in the area's average workweek series was relatively

44 Ann Arbor-Ypsilanti

CHART A 3
AVERAGE WEEKLY INITIAL CLAIMS FOR UNEMPLOYMENT INSURANCE, STATE PROGRAMS
ANN ARBOR-YPSILANTI SMSA
(Seasonally adjusted)

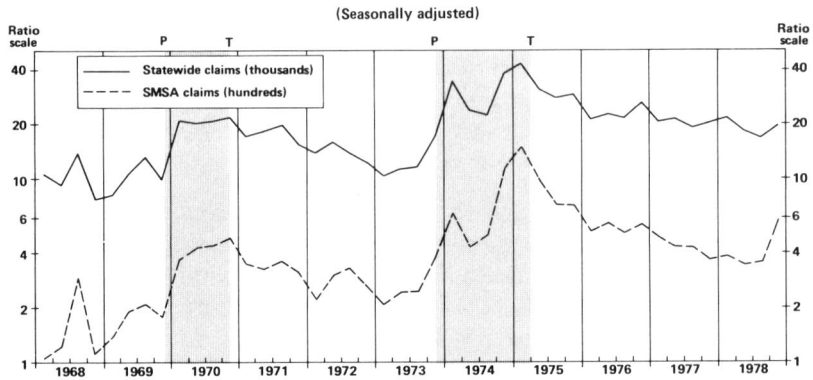

Source: The W.E. Upjohn Institute, based on data from the Michigan Employment Security Commission and the U.S. Department of Labor.
Shaded areas indicate national recession periods as defined by the National Bureau of Economic Research, Inc.
P = peak and T = trough.

CHART A 4
AVERAGE WEEKLY HOURS OF PRODUCTION WORKERS IN MANUFACTURING INDUSTRIES
ANN ARBOR — YPSILANTI SMSA
(Seasonally adjusted)

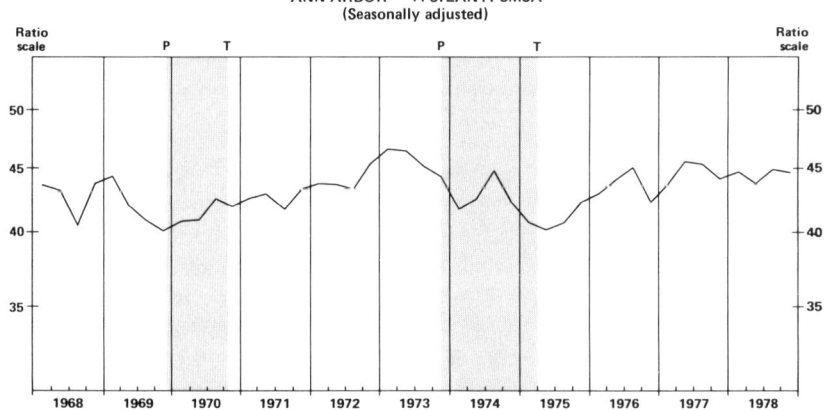

Source: Michigan Employment Security Commission.
Note: Seasonal adjustment by the W.E. Upjohn Institute.
Shaded areas indicate national recession periods as defined by the National Bureau of Economic Research, Inc.
P = peak and T = trough.

large, falling 9.5 percent during its 1969 contraction and 13.6 percent during its 1973-75 slump. The latter was exceeded only by the declines in the average workweek in the Flint and Lansing-East Lansing SMSAs. However, average weekly hours in the Ann Arbor-Ypsilanti SMSA never dropped below 40 during either downturn. Chart A4 reveals that the expansion in the average workweek since early 1975 has not brought it up to the pre-recession level of 46.5 hours, and that since mid-1977 a modest decline has occurred. Finally, the average workweek in this area exhibited virtually no secular trend over the 11-year period. The average annual growth rate for the series was estimated at only 0.4 percent. But, as Table A2 shows, this was above the growth rate for Michigan as a whole.

In summary, the behavior of several key employment and unemployment indicators in the Ann Arbor-Ypsilanti SMSA reveals a highly cyclically sensitive manufacturing sector, inducing relatively large cyclical swings in hours worked and employment in that sector. Moreover, these fluctuations contribute significantly to short-run swings in the area's jobless rate and initial claims for unemployment insurance. While the size of and relative stability in nonmanufacturing industries and the government sector in the area dampen the amplitude of cyclical swings in total wage and salary employment, overall labor market conditions in this metropolitan area were quite cyclically sensitive over the period examined.

Construction

Two series were used to measure construction activity in the metropolitan area—new building permits for private housing and employment in the local construction industry. They are shown in Charts A5 and A6, respectively.

New building permits for private housing reflect commitments to build and, nationally, peaks and troughs in this series tend to lead those of aggregate business activity. The local series is shown in Chart A5 in index form; that is, with each quarterly value expressed in terms of the series 1967 average to facilitate comparisons with the other metropolitan areas, the state, and the nation. The broken line represents the unadjusted quarterly index, which exhibits a considerable amount of erratic behavior. Nevertheless, cyclical upswings and downswings are apparent and are quite clear when a moving average of the unadjusted series is computed (shown by the solid line in Chart A5). Interestingly, the cyclical behavior of the local Index of New Building Permits did not conform to the national recession in 1969-70. Beginning in the third quarter of 1968, the local index (moving average) rose throughout the recession, reaching a cyclical high in the third quarter of 1972, one quarter

46 Ann Arbor-Ypsilanti SMSA

CHART A 5
INDEX OF NEW BUILDING PERMITS, PRIVATE HOUSING
ANN ARBOR - YPSILANTI SMSA
(1967 = 100)

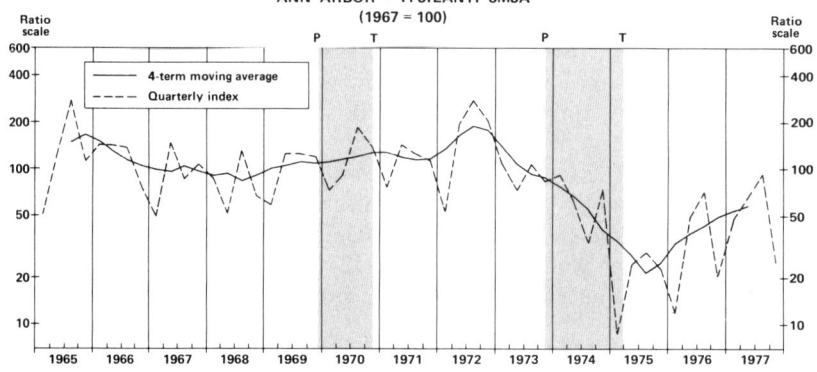

Source: The W.E. Upjohn Institute. Index based on the U.S. Department of Commerce, Bureau of the Census, *Construction Reports — Housing Authorized by Building Permits and Public Contracts, C-40*.
Shaded areas indicate national recession periods as defined by the National Bureau of Economic Research, Inc.
P = peak and T = trough.

CHART A 6
CONSTRUCTION EMPLOYMENT, ANN ARBOR-YPSILANTI SMSA
(Seasonally adjusted)

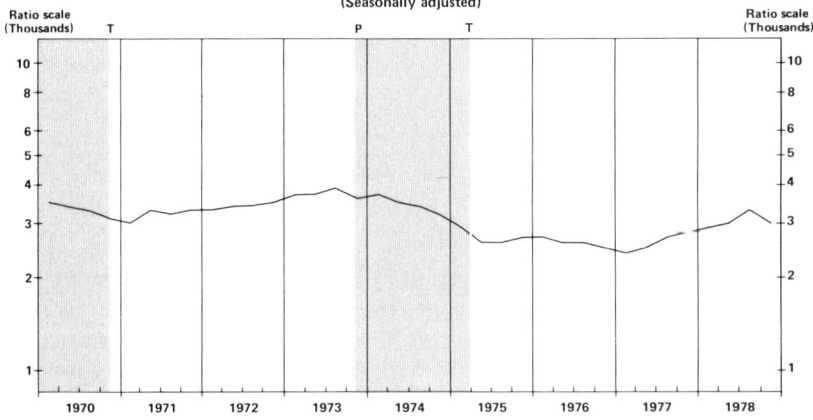

Source: Michigan Employment Security Commission.
Notes: Seasonal adjustment by the W.E. Upjohn Institute.
Shaded areas indicate national recession periods as defined by the National Bureau of Economic Research, Inc.
P = peak and T = trough.

before the national Index of New Building Permits and five quarters before the peak in aggregate business activity. The local downswing continued through the third quarter of 1975 and was of considerable magnitude. Measured from peak to trough, the index dropped 88.6 percent, exceeding the 69.1 percent decline registered by the national index and the 56.2 percent fall in the statewide index.[1] Moreover, the amplitude of the contraction in the Index of New Building Permits in this metropolitan area was the largest among the 11 metropolitan areas in Michigan. Chart A5 reveals that as of 1977 the Index of New Building Permits for the area remained considerably below pre-recession levels, and preliminary data for 1978 show that, although the index continued to rise, at 72.8 it was still far below the peak level of 193.5 registered in the third quarter of 1972.

Construction employment in the area, shown in Chart A6, also exhibits a high degree of cyclical sensitivity. Especially striking was the magnitude of the contraction (38.5 percent) which began after the third quarter of 1973 and continued through the first quarter of 1977. By the middle of 1978, construction employment had again expanded in the Ann Arbor-Ypsilanti SMSA, but had not yet come even close to the pre-recession level. For the 1970-78 period, the average annual growth rate for local construction employment was -2.8 percent, so that like new building permits for private housing, a negative trend characterized this series over the period examined.

Banking Activity

Several key banking indicators for the Ann Arbor-Ypsilanti SMSA outperformed their statewide counterparts over the 1970-78 period. Annual growth rates are shown in Table A3. In current-dollar terms, commercial bank deposits in the metropolitan area exhibited higher rates of growth over the 1970-77 period: 4.4 percent and 9.2 percent for demand and total deposits, respectively, in the metropolitan area, compared to 3.7 percent and 7.3 percent in the state as a whole. On a deflated basis—current-dollar values adjusted for price changes—demand deposits declined at a 2.4 percent annual rate in the Ann Arbor-Ypsilanti SMSA and at a 3.0 percent rate in the state. This metropolitan area also outperformed the state in terms of the growth of loans at commercial banks. Over

---

1. The national Index of New Building Permits is published monthly by the U.S. Department of Commerce, Bureau of Economic Analysis. It is a seasonally adjusted index (1967=100) so that it is not strictly comparable to the moving average of state data or of the local data shown in Chart A5.

the 1970-77 period, the current-dollar volume of loans expanded at a 10.2 percent annual rate in the metropolitan area, more than 2 percentage points above the state rate of 7.6 percent. And among the 11 metropolitan areas in Michigan, this area ranked third in the growth of commercial bank loans, exceeded only by the 10.3 percent annual growth rate in the Muskegon SMSA and the 11.6 percent growth rate in the Saginaw SMSA. Table A3 also shows that growth was relatively strong in two key loan categories in the Ann Arbor-Ypsilanti SMSA, commerical and industrial loans and consumer installment loans. The former grew at an annual rate higher than that of total loans.

As would be expected, deposits at commercial banks displayed significant cyclical movements which conformed to national business cycle patterns but with a lead. Chart A7 shows indexes of deflated demand and total deposits in the Ann Arbor-Ypsilanti SMSA; that is, the price-adjusted value of quarterly current-dollar totals expressed as a percent of the 1972 average.

TABLE A3

Average Annual Growth Rates of Selected
Commercial Banking Indicators
Ann Arbor-Ypsilanti SMSA and Michigan[a]

(percent)

Indicator	Ann Arbor-Ypsilanti SMSA	Michigan
Demand deposits (current dollars)	4.4	3.7
Deflated demand deposits[b]	-2.4	-3.0
Total deposits (current dollars)	9.2	7.3
Deflated total deposits[b]	2.2	0.4
Total loans (current dollars)	10.2	7.6
Commercial and industrial loans (current dollars)	13.3	7.3
Consumer installment loans (current dollars)	9.9	9.1

[a] Except where indicated otherwise, estimated growth rates are based on log-linear trends for the 1970-78 period.

[b] Current-dollar values adjusted for changes in the U.S. Consumer Price Index.

Both indexes began their cyclical downswings before the start of the 1973-75 recession. Thus, peaks in both deposit indexes preceded

Ann Arbor-Ypsilanti SMSA

CHART A 7
INDEX OF DEFLATED TOTAL DEPOSITS AND INDEX OF DEFLATED DEMAND DEPOSITS
ANN ARBOR - YPSILANTI SMSA
(1972 = 100)
(Seasonally adjusted)

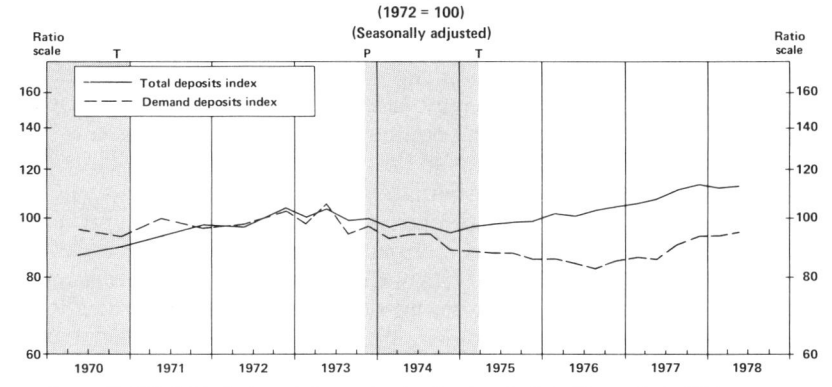

Source: The W.E. Upjohn Institute. Indices are based on data from the Federal Reserve Bank of Chicago.
Note: Shaded areas indicate national recession periods as defined by the National Bureau of Economic Research, Inc. P = peak and T = trough.

CHART A 8
COMMERCIAL BANK LOANS, ANN ARBOR - YPSILANTI SMSA
(Current dollars)

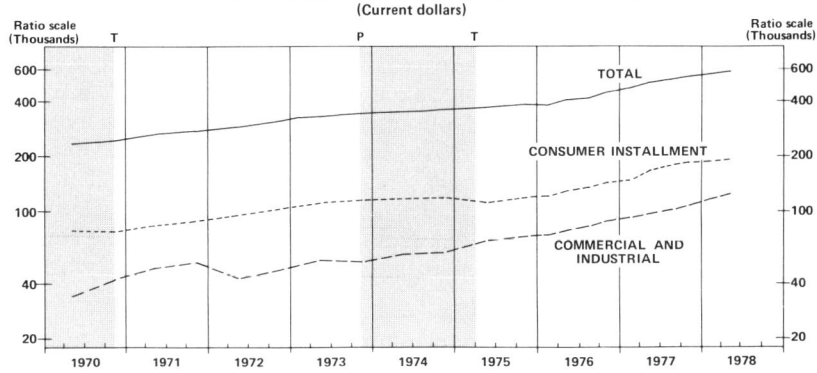

Source: Federal Reserve Bank of Chicago.
Notes: Seasonal adjustment of total loans by the W.E. Upjohn Institute. Other loans are not seasonally adjusted.
Shaded areas indicate national recession periods as defined by the National Bureau of Economic Research, Inc. P = peak and T = trough.

those of total wage and salary and manufacturing employment in the area. However, while the trough in the Index of Deflated Total Deposits occurred one quarter ahead of national aggregate business activity, the Index of Deflated Demand Deposits lagged considerably. This lag in demand deposits was also evident in the other metropolitan areas and in the state as a whole. Moreover, the real value of the national money stock, M1, which is composed of cash and demand deposits at commercial banks, also lagged at the 1975 trough, and its expansion so far has been quite weak. Thus, the behavior of the local Index of Deflated Demand Deposits during the recent expansion is not particularly unusual. The relatively slow growth of demand deposits compared to the rate of inflation constrained the expansions of the national money supply series, M1, and the deflated values of demand deposits at the state and local level. Additionally, technological changes in banking—electronic transfer of funds, for example—may result in less vigorous increases in demand deposits compared to total deposits which include savings accounts.

The magnitudes of contractions and expansions in the two local deposit indexes are substantially different. The Index of Deflated Demand Deposits fell 21.6 percent from peak to trough over the 1973-76 period. The Index of Deflated Total Deposits, on the other hand, declined 8.4 percent over a shorter period. It should be noted that, for both indexes, the cyclical contraction in the Ann Arbor-Ypsilanti SMSA was less than that of the state (31.1 percent and 11.2 percent, respectively) and, except for the Saginaw SMSA, the mildest among Michigan metropolitan areas. Since hitting a cyclical low in the third quarter of 1976, the Index of Deflated Demand Deposits has risen 14.8 percent but, as Chart A7 reveals, it is still far below pre-recession levels. In contrast, the Index of Deflated Total Deposits rose 19 percent from its cyclical low in the fourth quarter of 1974, and its most recent levels are considerably above the pre-recession peak.

Total loans, commercial and industrial loans, and consumer installment loans are shown in Chart A8. The current-dollar value of loans in the Ann Arbor-Ypsilanti SMSA was quite cyclically stable, with no absolute decline evident during the 1973-75 recession. A slowdown in the growth rate occurred, but it was shortlived. As Chart A8 shows, total loans in the area have increased considerably since 1975. Commercial and industrial loans generally grew throughout the 1970-78 period, exhibiting no cyclical sensitivity. Consumer installment loans displayed some cyclical behavior, but this was brief and mild. Therefore, loan activity at commercial banks in the Ann Arbor-Ypsilanti SMSA was characterized by relatively steady growth in current-dollar volume despite the severe recession that affected the nation in the 1973-75 period.

Battle Creek SMSA

The Battle Creek SMSA is Michigan's most recently designated metropolitan area. That designation occurred in 1971 when Calhoun County was defined as a Standard Metropolitan Statistical Area, with Battle Creek as the central city. Barry County was added in 1973 so that presently the Battle Creek SMSA encompasses a two-county area. One of Michigan's smaller metropolitan areas, it ranks eighth in population among the 11 metropolitan areas. In 1970, its population was 180,129, or 2 percent of the state total. The area's population rose 1.0 percent between 1970 and 1977, second lowest among Michigan metropolitan areas, and was below the statewide population rise of 2.8 percent over the seven-year period.

The Battle Creek SMSA had a total personal income level of $1,184 million in 1976 and a per capita income level of $6,491. The latter ranked sixth among Michigan SMSAs and was below the statewide level of $6,757 and the all-U.S. SMSAs' average of $6,824. Over the 1969-76 period, total personal income in the Battle Creek SMSA rose at an average annual rate of 8.8 percent, sixth among the 11 Michigan SMSAs, and below the nationwide rate of 9.3 percent and the all-U.S. SMSA rate of 9.0 percent. Since 1969, the Battle Creek SMSA has been characterized by slow population growth and modest growth in personal income.

The Battle Creek SMSA is more industrialized than the state as a whole (see Table BC1). In 1972, 40.8 percent of its total wage and salary employment was accounted for by manufacturing firms. By 1977, however, the manufacturing share had fallen to 36.0 percent, as nonmanufacturing and government employment increased in relative size. A major difference between this area's industrial structure and that of the state is that manufacturing employment is more evenly divided between durable and nondurable goods industries in the local area. The latter accounted for 15.9 percent and 15.5 percent of total wage and salary employment in the area in 1972 and 1977, respectively. Statewide, the nondurable goods industries' share of total wage and salary employment was 7.0 percent in 1972 and 6.5 percent in 1977, less than one-half the proportion in the Battle Creek SMSA.

TABLE BC1

Percentage Distribution of Total Wage and Salary Employment
Battle Creek SMSA and Michigan, 1972 and 1977

Item	Battle Creek SMSA		Michigan	
	1972	1977	1972	1977
Total Wage and Salary	100.0%	100.0%	100.0%	100.0%
Manufacturing	40.8	36.0	35.1	32.4
Durables	24.9	20.6	28.1	25.9
Nondurables	15.9	15.5	7.0	6.5
Nonmanufacturing	41.7	44.7	48.0	49.6
Government	17.5	19.3	16.9	18.0

Source: Michigan Employment Security Commission.

The food and kindred products industry accounts for the largest share of manufacturing workers in the area. In 1977, for example, 32.2 percent of local manufacturing workers were employed in that industry. The area's second largest industry in terms of employment is nonelectrical machinery, which accounted for 31.8 percent of local manufacturing workers in 1977. The Battle Creek SMSA has only a small proportion of manufacturing workers engaged in the production of transportation equipment. In 1977, for example, that industry accounted for only 0.9 percent of the area's manufacturing workers, the smallest share among the 11 SMSAs in Michigan and considerably below the high of 63.8 percent in the Lansing-East Lansing SMSA. Taken by itself, the industrial structure of the Battle Creek SMSA suggests that the area is probably less cyclically sensitive than the state as a whole, at least in terms of employment. The following sections show that this is indeed the case for local employment indicators, but local construction and banking activity have exhibited a fairly high degree of sensitivity to cyclical downturns in the national economy.

Business Conditions in the Battle Creek SMSA

Labor Market Conditions

Chart BC1 shows four major employment categories, by quarter, over the 1970-78 period. Manufacturing employment was the most cyclically sensitive, falling 15.7 percent from its peak in the first quarter of 1973 to its low in the second quarter of 1975. While the contraction in the area's manufacturing employment was milder than the 18.7 percent decline recorded statewide, it was of somewhat longer duration, lasting nine quarters compared to five quarters for Michigan as a whole. Besides, the length of the cyclical decline in the area's manufacturing employment was exceeded only by the twelve-quarter contraction in the adjacent Jackson SMSA.

The recovery in manufacturing employment in the Battle Creek SMSA was quite weak, however. After hitting a cyclical low in the second quarter of 1975, manufacturing employment had risen just under 5 percent by the end of 1978. Chart BC1 shows that this level was considerably below the pre-recession peak.

The area's nonmanufacturing employment also fell during the recession, but only a mild 2.2 percent. During the current expansion it increased 15.5 percent, and as of the fourth quarter of 1978 local nonmanufacturing employment stood almost 13 percent above its previous peak.

Chart BC1 shows that government employment experienced a brief decline during the 1973-75 recession. Despite the brevity, however, the relative decline in public sector employment was the largest among the state's metropolitan areas. In fact, eight of the eleven SMSAs experienced little or no absolute loss of government employment during the last recession. Many actually registered increases.

Therefore, the 6.3 percent decline in this area's total wage and salary employment during the last recession was due to job losses in each of the three major categories shown in Chart BC1. That contraction was the sixth most severe among the 11 SMSAs, despite the relatively modest decline in manufacturing employment. The expansion in the area's total wage and salary employment— 10.5 percent since the recession—resulted largely from increases in the nonmanufacturing and government sectors. As already noted, manufacturing employment in the area has increased only slightly in the last three years.

54   Battle Creek SMSA

CHART BC 1
WAGE AND SALARY EMPLOYMENT, BATTLE CREEK SMSA
(Seasonally adjusted)

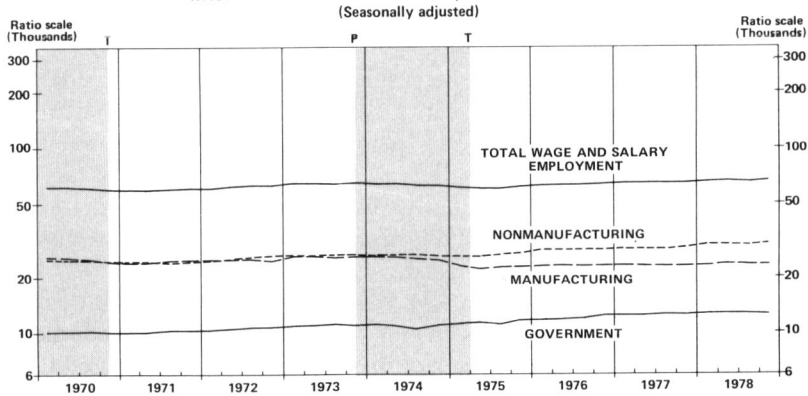

Source: Michigan Employment Security Commission.
Notes: Seasonal adjustment by the W.E. Upjohn Institute.
Shaded areas indicate national recession periods as defined by the National Bureau of Economic Research, Inc.
P = peak and T = trough.

CHART BC 2
UNEMPLOYMENT RATE
BATTLE CREEK SMSA
(Seasonally adjusted)

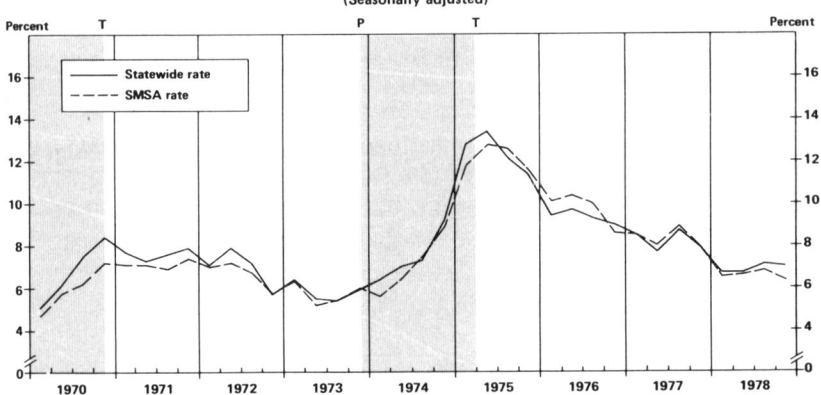

Source: Michigan Employment Security Commission.
Note: Seasonal adjustment by the W.E. Upjohn Institute.
Shaded areas indicate national recession periods as defined by the National Bureau of Economic Research, Inc.
P = peak and T = trough.

Table BC2 gives the annual growth rates for selected labor market indicators in the Battle Creek SMSA. Each of the three major employment categories—manufacturing, nonmanufacturing, and government—grew more slowly in the Battle Creek SMSA, compared to Michigan as a whole over the 1970-78 period. And the area's manufacturing employment exhibited negative growth. Except for the annual decline of 1.5 percent in the Jackson SMSA, the secular downtrend of manufacturing employment in the Battle Creek SMSA represents the poorest performance among the state's metropolitan areas over this nine-year period. Combined with the relatively slow growth of the area's nonmanufacturing and government sectors, this downtrend in manufacturing employment resulted in a growth rate for total wage and salary employment of only 0.9 percent, the slowest growth among Michigan metropolitan areas.

TABLE BC2

Annual Growth Rates of Selected Labor Market Indicators
Battle Creek SMSA and Michigan[a]

(percent)

Indicator	Battle Creek SMSA	Michigan
Total wage and salary employment	0.9	1.8
Manufacturing employment	-1.4	0.2
Nonmanufacturing employment	2.2	2.6
Government employment	2.7	3.0
Civilian labor force	0.9	2.0
Unemployment rate	4.0	2.8
Average weekly initial claims for UI[b]	10.0	7.3
Average workweek, production workers, mfg.	0.2	0.6

[a]Except where indicated otherwise, estimated growth rates are based on log-linear trends for the 1970-78 period.

[b]Computed for the 1968-78 period.

56   Battle Creek SMSA

The area's unemployment rate is shown in Chart BC2. The cyclical pattern of the local jobless rate is very similar to the state as a whole. In the Battle Creek SMSA, the jobless rate rose 146.2 percent from a cyclical low in the spring of 1973 to a high in the second quarter of 1975. That cyclical upswing was about equal to the statewide rise of 148.2 percent, and was the second lowest among the 11 metropolitan areas. Interestingly, the Detroit SMSA exhibited the smallest amplitude in the unemployment rate during the last recession. At the bottom of the last recession, the jobless rate in the Battle Creek SMSA was 12.8 percent, compared to 13.4 percent for Michigan as a whole.

As Chart BC2 shows, both the local and state jobless rates have fallen substantially during the recent business expansion. However, at the end of 1978 both were still above pre-recession lows. Overall, the jobless rate in the Battle Creek SMSA is less cyclically sensitive than in the state as a whole, but not considerably so.

Over the entire 1970-78 period, however, the jobless rate in the Battle Creek SMSA did drift upward at a more rapid rate than the state, 4 percent compared to 2.8 percent (see Table BC2). This local uptrend was exceeded only by the 4.3 percent growth rate in the Jackson SMSA. Table BC2 shows that labor-force growth in the Battle Creek SMSA was not rapid, falling below the state rate. In addition, the 0.9 percent growth rate in the area's labor force over the nine-year period represents the slowest growth among the 11 metropolitan areas.

Average weekly initial claims for unemployment insurance, shown in Chart BC3, moved up somewhat more during the 1973-75 recession and fell somewhat less on average during expansions in the Battle Creek SMSA, compared to the state as a whole. Over the nine-year period, initial claims also grew at an annual rate of 10.0 percent in the local area, second highest among Michigan metropolitan areas and substantially above the state rate of 7.3 percent. As Chart BC3 shows, average weekly initial claims for unemployment insurance in the Battle Creek SMSA in 1978 remained well above the previous cyclical low in 1973.

The average workweek of manufacturing production workers is relatively stable in the Battle Creek SMSA. Chart BC4 shows that cyclical swings in this local series conformed closely to national business cycle patterns, declining during recessions and increasing during business expansions, but with a slight leading tendency. Over the 1970-78 period, the average workweek in the area varied

Battle Creek SMSA   57

CHART BC 3
AVERAGE WEEKLY INITIAL CLAIMS FOR UNEMPLOYMENT INSURANCE, STATE PROGRAMS
BATTLE CREEK SMSA
(Seasonally adjusted)

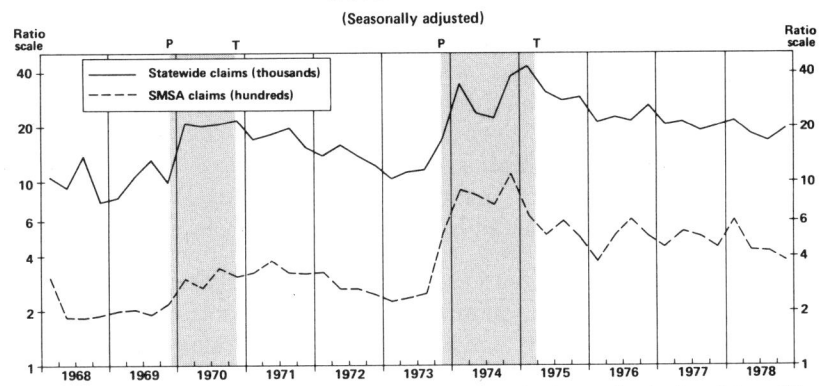

Source: The W.E. Upjohn Institute, based on data from the Michigan Employment Security Commission and the U.S. Department of Labor.
Shaded areas indicate national recession periods as defined by the National Bureau of Economic Research, Inc.
P = peak and T = trough.

CHART BC 4
AVERAGE WEEKLY HOURS OF PRODUCTION WORKERS IN MANUFACTURING INDUSTRIES
BATTLE CREEK SMSA
(Seasonally adjusted)

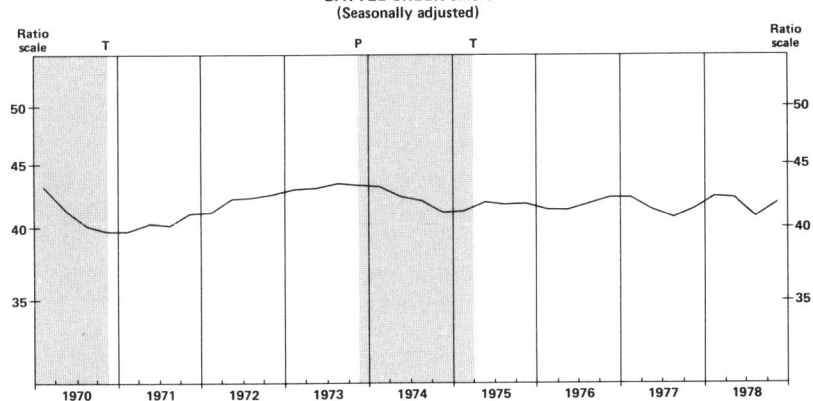

Source: Michigan Employment Security Commission.
Note: Seasonal adjustment by the W.E. Upjohn Institute.
Shaded areas indicate national recession periods as defined by the National Bureau of Economic Research, Inc.
P = peak and T = trough.

between a low of 39.7 hours recorded in the last quarter of 1970 and a high of 43.4 hours registered in the third quarter of 1973. That is a relatively narrow range compared to the wide swings in average weekly hours in the Flint, Lansing-East Lansing, and Saginaw SMSAs. The latter have manufacturing sectors more heavily concentrated in durable goods production, primarily related to automobiles. In the Battle Creek SMSA, production of nondurable goods—cereals, for example—accounts for a larger share of total manufacturing than in most of the state's other SMSAs, particularly those dominated by automobile and related production.

The industrial structure of the Battle Creek SMSA, which is more heavily dependent on nondurable goods production than other Michigan metropolitan areas except the Kalamazoo-Portage SMSA, appears to account for the area's relative cyclical stability, especially in terms of employment and hours worked in manufacturing. But this type of industrial concentration has also generated slow growth in the area (negative growth for manufacturing employment) and has not improved the area's unemployment situation, as indicated by the behavior of the local jobless rate and initial claims for unemployment insurance. If anything, the local industrial structure appears to have exacerbated the unemployment situation over the period examined.

Construction

Two measures of construction activity, new building permits for private housing and construction employment, are shown in Charts BC5 and BC6, respectively. The former shows two series of building permit data for the Battle Creek SMSA—a quarterly unadjusted series (dashed line) and a moving average of the unadjusted series (solid line)—both given in index form with 1967=100.

The unadjusted Index of New Building Permits displayed a high degree of erratic behavior over the 1965-77 period, largely masking any cyclical movements. The smoother index based on the moving average reduces the erratic behavior and reveals cyclical swings that do generally conform to national business cycles. Note that the moving average index declined during both recessions and rose, for the most part, during expansions. Not surprisingly, Chart BC5 shows that the cyclical downswing in new building permits in the area was less severe during the 1969-70 period than during the 1973-75 recession. Measured from peak to trough, the index fell 53.4 percent during the former period and 84.5 percent during the latter.

Battle Creek SMSA 59

CHART BC 5
INDEX OF NEW BUILDING PERMITS, PRIVATE HOUSING
BATTLE CREEK SMSA
(1967 = 100)

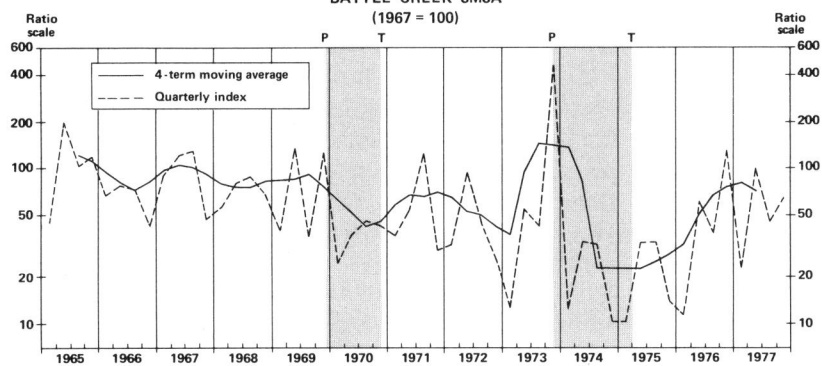

Source: The W.E. Upjohn Institute. Index based on the U.S. Department of Commerce, Bureau of the Census, *Construction Reports — Housing Authorized by Building Permits and Public Contracts, C-40.*
Shaded areas indicate national recession periods as defined by the National Bureau of Economic Research, Inc.
P = peak and T = trough.

CHART BC 6
CONSTRUCTION EMPLOYMENT, BATTLE CREEK SMSA
(Seasonally adjusted)

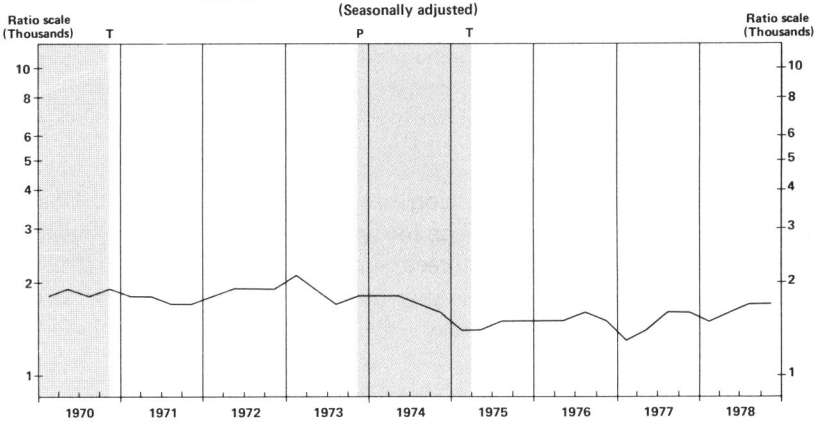

Source: Michigan Employment Security Commission.
Notes: Seasonal adjustment by the W.E. Upjohn Institute.
Shaded areas indicate national recession periods as defined by the National Bureau of Economic Research, Inc.
P = peak and T = trough.

These local contractions exceeded the statewide declines of 21.6 percent and 56.2 percent for the 1969-70 and 1973-75 recessions, respectively. Chart BC5 shows that, despite the steady rise during the current expansion, the Index of New Building Permits in the Battle Creek SMSA was still below its pre-recession peak. Preliminary data for 1978 show index levels below that of the first quarter of 1977, the highest level achieved to date during the expansion period.

Total construction employment in the Battle Creek SMSA also exhibited a high degree of cyclical sensitivity. Chart BC6 shows that the peak in construction employment led the national business cycle peak of late 1973 by several quarters and lagged at both troughs. During the last recession, construction employment fell 33.3 percent from its high in the first quarter of 1975, which exceeds the statewide decline of 23.5 percent. Among the state's 11 metropolitan areas, the contraction in the Battle Creek SMSA ranked fifth in terms of severity. Chart BC5 also reveals that during the current expansion, the upswing in local construction employment has been weak. As of the last quarter of 1978, employment was just 21.4 percent above its cyclical low of the second quarter of 1975. The relatively large cyclical decline during the mid-1970s and the weak expansion in the last three years impart an overall downward trend to construction employment in the Battle Creek SMSA. For the 1970-78 period annual growth was -2.5 percent.

Banking Activity

Because of its relatively recent designation as a metropolitan area, commercial banking data for the Battle Creek SMSA are available starting with the fourth quarter of 1971. (Data shown in this study start with the second quarter of 1970 for the other SMSAs in the state.) This eliminates any measurement of behavior during the 1969-70 recession. Furthermore, the number of banks reporting to the Federal Reserve doubled from three to six between the first and second quarters of 1973. As a result, the loans and deposits series display an abrupt jump at that time. However, beginning with the second quarter of 1973 the banking series are continuous, allowing measurement of recession behavior in the 1973-75 period and during the subsequent expansion.

An Index of Deflated Demand Deposits—the price-adjusted values of current-dollar levels expressed relative to the 1972 average—is shown in Chart BC7. This local index started declining before the 1973-75 recession began and continued its cyclical downswing through the third quarter of 1976. The magnitude of that decline was 28.9 percent, second highest among the 11 metropolitan areas in Michigan. The broader Index of Deflated Total Deposits, also shown in Chart BC7, fell a more moderate 15.7 percent from its cyclical high in early 1974 to a low in the first quarter of 1976. The lag in both indexes is not unique to the Battle Creek SMSA, since each metropolitan area and the state as a whole exhibited lags at the 1975 trough.

Chart BC7 shows that both indexes were still well below their previous peaks as of the second quarter of 1978, the latest quarter for which data were available. The Index of Deflated Demand Deposits rose 19 percent from its cyclical low, and the Index of Deflated Total Deposits was up 9.7 percent.

Annual growth rates for the 1973-77 period are given in Table BC3. In current-dollar terms, demand deposits and total deposits at commercial banks in the Battle Creek SMSA exhibited growth exceeding the statewide rates. On a deflated basis, however, negative growth characterized these series at the state and local levels.

Commercial bank loans in the Battle Creek SMSA are shown in Chart BC8. On a current-dollar basis, total loans fell during the 1973-75 period, as did consumer installment loans and commercial and industrial loans. The current-dollar volume of total loans declined 9.8 percent from its cyclical peak in the first quarter of 1974 to a trough two years later. Among Michigan metropolitan areas, that was the most severe contraction. But total loans in the area recovered during 1976 and 1977 and in the second quarter of 1978 had risen 41.7 percent, standing 27.8 percent above the 1974 peak. For the 1973-77 period, the current-dollar volume of total loans grew at an annual rate of 3.4 percent, which is below the growth rate for the state over the same period (see Table BC3).

During the 1973-75 recession, consumer installment loans fell 32.4 percent and commercial and industrial loans declined 26.2 percent in the Battle Creek SMSA, both ranking first among the state's metropolitan areas. As Chart BC8 shows, commercial and industrial loans rose to a level in the third quarter of 1977 that exceeded their previous peak, but have fallen since that time. Consumer installment loans, on the other hand, have only recently moved above their 1973 peak, rising 49.7 percent from a cyclical low in the second quarter of 1976.

62 Battle Creek SMSA

CHART BC 7
INDEX OF DEFLATED TOTAL DEPOSITS AND INDEX OF DEFLATED DEMAND DEPOSITS
BATTLE CREEK SMSA
(1972 = 100)
(Seasonally adjusted)

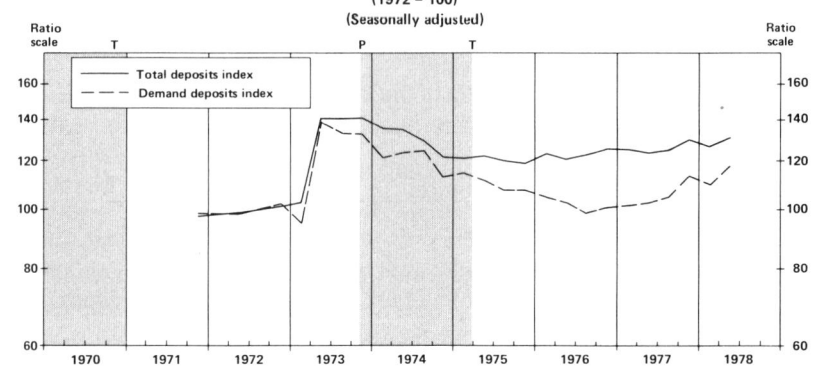

Source: The W.E. Upjohn Institute. Indices are based on data from the Federal Reserve Bank of Chicago.
Note: Shaded areas indicate national recession periods as defined by the National Bureau of Economic Research, Inc. P = peak and T = trough.

Battle Creek was declared an SMSA in 1971. Federal Reserve Board data are not available prior to that date.

CHART BC 8
COMMERCIAL BANK LOANS, BATTLE CREEK SMSA
(Current dollars)

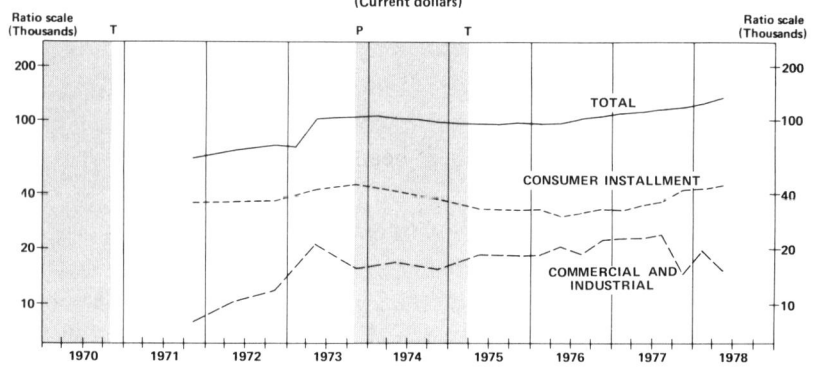

Source: Federal Reserve Bank of Chicago.
Notes: Seasonal adjustment of total loans by the W.E. Upjohn Institute. Other loans are not seasonally adjusted.
Shaded areas indicate national recession periods as defined by the National Bureau of Economic Research, Inc. P = peak and T = trough.

TABLE BC3

Annual Growth Rates of Selected Commercial Banking Indicators Battle Creek SMSA and Michigan[a]

(percent)

Indicator	Battle Creek SMSA	Michigan
Demand deposits (current dollars)	2.6	1.5
Deflated demand deposits[b]	-4.9	-5.7
Total deposits (current dollars)	6.4	6.3
Deflated total deposits[b]	-1.4	-1.2
Total loans (current dollars)	3.4	4.9
Commercial and industrial loans (current dollars)	4.7	6.7
Consumer installment loans (current dollars)	-5.6	6.0

[a] Estimated growth rates are based on log-linear trends for the 1973-77 period.

[b] Current-dollar values adjusted for changes in the U.S. Consumer Price Index.

The overall performance of loans in the Battle Creek SMSA has been weak compared to other metropolitan areas in Michigan. Consumer installment loans have been a particularly poor performer in this area since 1973.

Bay City SMSA

The Bay City SMSA is the smallest metropolitan area in Michigan. It consists of a single county—Bay County—with Bay City as its urban center. In 1970, this metropolitan area had a population of 117,339, which represented 1.3 percent of the state total. By 1977, the population of the Bay City SMSA had risen 2.5 percent, but its share of the total state population was unchanged. Over this period, growth in the area's population ranked seventh among the 11 metropolitan areas in Michigan.

Total personal income in the Bay City SMSA amounted to $730 million in 1976, up 79.8 percent from the 1969 level of $406 million. Over that period, the area registered an 8.7 percent annual growth rate of personal income, which is only slightly below the 9.0 percent growth rate for all U.S. SMSAs. Among Michigan metropolitan areas, however, this area's personal income growth ranked only seventh, and its per capita income was second lowest at $6,116 in 1976. For that year, only the Muskegon SMSA had a lower per capita income level.

Table BY1 shows that the employment distribution in the Bay City SMSA is very similar to that of the state, with a slightly smaller share engaged in manufacturing and a slightly larger share accounted for by nonmanufacturing industries. Over one-half of all wage and salary workers in the Bay City SMSA work in nonmanufacturing activities, the highest proportion among the 11 metropolitan areas in the state.

Like Michigan as a whole, the bulk of manufacturing workers in the Bay City SMSA are employed by durable goods producing industries. This metropolitan area is only slightly less concentrated in durable goods production than is Michigan as a whole. Within the manufacturing sector, however, a relatively large proportion of

TABLE BY1

Percentage Distribution of Total Wage and Salary Employment Bay City SMSA and Michigan, 1972 and 1977

Item	Bay City SMSA 1972	Bay City SMSA 1977	Michigan 1972	Michigan 1977
Total Wage and Salary	100.0%	100.0%	100.0%	100.0%
Manufacturing	31.5	30.5	35.1	32.4
Durables	25.2	23.3	28.1	25.9
Nondurables	6.2	7.2	7.0	6.5
Nonmanufacturing	53.8	54.8	48.0	49.6
Government	14.8	14.7	16.9	18.0

Source: Michigan Employment Security Commission.

workers in the Bay City SMSA are employed by the transportation equipment industry. In 1977, about 40 percent of the area's production workers were employed in that industry, compared to 34.6 percent for the state as a whole.

Overall, the Bay City SMSA has an industrial structure that is not considerably different from that of the state. Over roughly the last ten years, however, this small metropolitan area was characterized by slower population and income growth than the state as a whole.

Business Conditions in the Bay City SMSA

Labor Market Conditions

The behavior of wage and salary employment in the Bay City SMSA is shown in Chart BY1 for the 1970-78 period. Over those nine years manufacturing employment grew at an annual rate of 1.1 percent, exceeding the 0.2 percent growth rate for the state as a whole (see Table BY2). Nonmanufacturing employment grew at an annual rate of 3.5 percent, which is also above the statewide rate (2.6 percent). Among Michigan metropolitan areas, the growth of nonmanufacturing employment in the Bay City SMSA ranked third behind the 5.2 percent and 4.6 percent rates of growth in the Ann Arbor-Ypsilanti SMSA and Kalamazoo-Portage SMSA, respectively. However, government employment in this area grew at a relatively slow pace of 2.2 percent, fourth lowest among the state's 11 metropolitan areas and below the 3.0 percent growth rate for the state as a whole.

The relatively strong growth in employment in the area's manufacturing and nonmanufacturing industries resulted in a 2.5 percent annual growth rate for total wage and salary employment. This ranked fourth among the 11 metropolitan areas over the 1970-78 period and, as Table BY2 shows, exceeded the growth statewide.

Chart BY1 reveals that, in general, the behavior of total wage and salary, manufacturing, and nonmanufacturing employment conformed to national business cycle patterns, exhibiting upswings

TABLE BY2

Annual Growth Rates of Selected Labor Market Indicators
Bay City SMSA and Michigan[a]

(percent)

Indicator	Bay City SMSA	Michigan
Total wage and salary employment	2.5	1.8
Manufacturing employment	1.1	0.2
Nonmanufacturing employment	3.5	2.6
Government employment	2.2	3.0
Civilian labor force	1.9	2.0
Unemployment rate	-1.0	2.8
Average weekly initial claims for UI[b]	3.1	7.3
Average workweek, production workers, mfg.[b]	1.3	0.2

[a] Except where indicated otherwise, estimated growth rates are based on log-linear trends for the 1970-78 period.

[b] Computed for the 1968-78 period.

and downswings that began at about the same time as those of aggregate business activity nationwide. From late 1970 through the fourth quarter of 1973, total wage and salary employment rose 19.8 percent, with manufacturing employment increasing 20.9 percent, nonmanufacturing 23.0 percent, and government employment a comparatively modest 9.3 percent. But as Chart BY1 shows, the latter continued to expand throughout the 1973-75 recession while each of the others declined, manufacturing and nonmanufacturing employment by 19.2 percent and 2.2 percent, respectively, and total wage and salary employment by 6.0 percent. Six of the other metropolitan areas in the state experienced more severe contractions in total wage and salary employment than the Bay City SMSA.

From the trough of the recession in the first quarter of 1975 to the fourth quarter of 1978, the area's total wage and salary employment rose 19.5 percent, with manufacturing employment increasing the most, 39.3 percent. Over that period, nonmanufacturing and government employment expanded 13.9 percent and 8.2 percent, respectively.

Thus, as is the case for the other metropolitan areas in Michigan, the relative cyclical stability of the nonmanufacturing and government sectors offsets, to some degree, the cyclical sensitivity of local manufacturing industries. As noted above, the Bay City SMSA has a manufacturing sector heavily concentrated in durable goods production. In 1977, the proportion of total wage and salary workers engaged in such production was only slightly less in this area than statewide—23.3 percent and 25.9 percent, respectively. While the cyclical behavior of employment in the Bay City SMSA and the state as a whole was not markedly different over the 1970-78 period, the local area did experience a somewhat greater growth of employment.

The unemployment rates for the Bay City SMSA and Michigan are shown by quarter in Chart BY2. Except for the 1970-72 years when the local unemployment rate was well above the state average, the two have been quite close in terms of level and direction. In the Bay City SMSA, the jobless rate moved up to a higher level than in the state during the last recession, reaching a seasonally adjusted 14.3 percent in the first quarter of 1975. That was 150.9 percent above the low in the third quarter of 1973 and very close to the 148.1 percent rise in the statewide average. As Chart BY2 reveals,

Bay City SMSA 69

CHART BY 1
WAGE AND SALARY EMPLOYMENT, BAY CITY SMSA
(Seasonally adjusted)

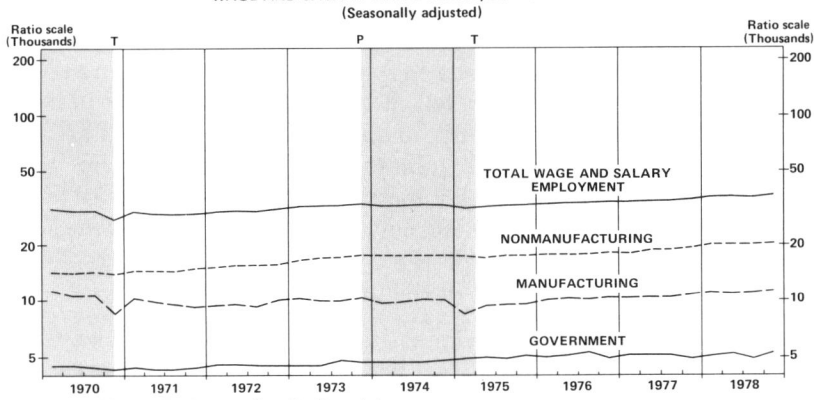

Source: Michigan Employment Security Commission.
Notes: Seasonal adjustment by the W.E. Upjohn Institute.
Shaded areas indicate national recession periods as defined by the National Bureau of Economic Research, Inc.
P = peak and T = trough.

CHART BY 2
UNEMPLOYMENT RATE
BAY CITY SMSA
(Seasonally adjusted)

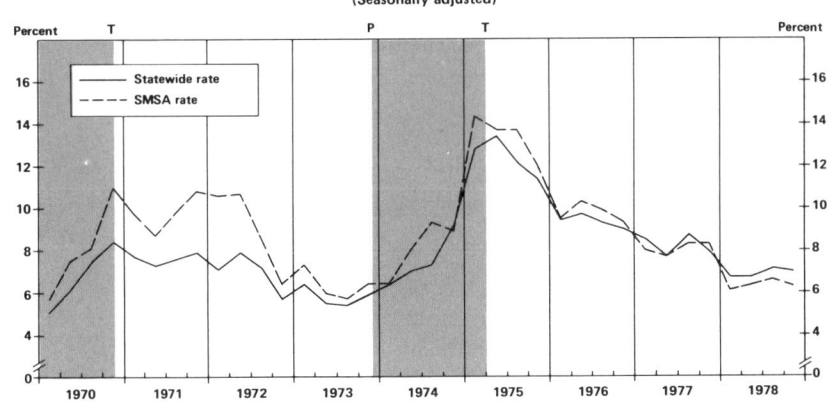

Source: Michigan Employment Security Commission.
Note: Seasonal adjustment by the W.E. Upjohn Institute.
Shaded areas indicate national recession periods as defined by the National Bureau of Economic Research, Inc.
P = peak and T = trough.

the behavior of this area's unemployment rate during the current expansion has been very close to the state's performance. However, in 1978, the jobless rate in the Bay City SMSA remained below the average for Michigan as a whole. Over the entire nine-year period, the jobless rate in this area fell at an annual rate of 1.0 percent compared to a 2.8 percent annual increase statewide. The downtrend in this area's jobless rate represents the best performance among Michigan metropolitan areas from 1970 to 1978. Except for the Grand Rapids SMSA, which experienced a modest downtrend of -0.6 percent, all other metropolitan areas experienced an upward drift in the unemployment rate.

Average weekly initial claims for unemployment insurance, shown in Chart BY3, also reveal behavior in the local area that is similar to the state, both series beginning cyclical upswings before peaks in national business activity and cyclical downswings at about the same time a national expansion starts. In relative terms, initial claims rose more in the Bay City SMSA than in the state during the 1969-70 recession (324.8 percent compared to 193.2 percent), but during the last recession the relative increases were 301.7 percent and 301.1 percent for the local area and state, respectively. As of the fourth quarter of 1978, average weekly initial claims in the Bay City SMSA were 65.8 percent below the recession high of the first quarter of 1975. Over the same period, initial claims statewide were down 53.7 percent. Thus, like the unemployment rate, initial claims in the Bay City SMSA have outperformed the state series during the current expansion. Furthermore, over the 1968-78 period, initial claims for unemployment insurance grew at an annual rate of 3.1 percent in the Bay City area compared to a 7.3 percent growth rate for Michigan as a whole, the third lowest among the state's 11 metropolitan areas.

The average workweek of manufacturing workers in the Bay City SMSA is shown by quarter for the 1968-78 period in Chart BY4. Unlike all other metropolitan areas in Michigan, the average workweek in the Bay City SMSA rose throughout most of the 1973-75 recession. Statewide, average weekly hours of manufacturing workers fell 9.7 percent during that period, with substantial declines recorded in the Lansing-East Lansing and Flint SMSAs. The rising workweek in the manufacturing sector of this local area was certainly atypical behavior for a Michigan metropolitan area. Over the entire period, the average workweek in the area grew at an annual rate of 1.3 percent compared to almost no growth statewide and a slight downtrend nationwide.

Bay City SMSA 71

CHART BY 3
AVERAGE WEEKLY INITIAL CLAIMS FOR UNEMPLOYMENT INSURANCE, STATE PROGRAMS
BAY CITY SMSA
(Seasonally adjusted)

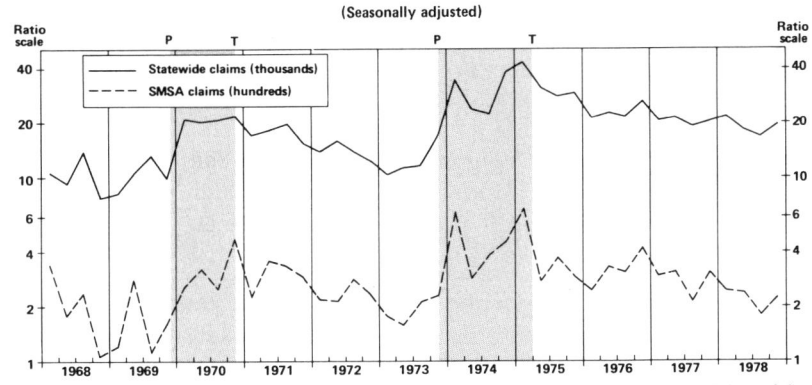

—— Statewide claims (thousands)
- - - SMSA claims (hundreds)

Source: The W.E. Upjohn Institute, based on data from the Michigan Employment Security Commission and the U.S. Department of Labor.
Shaded areas indicate national recession periods as defined by the National Bureau of Economic Research, Inc.
P = peak and T = trough.

CHART BY 4
AVERAGE WEEKLY HOURS OF PRODUCTION WORKERS IN MANUFACTURING INDUSTRIES
BAY CITY SMSA
(Seasonally adjusted)

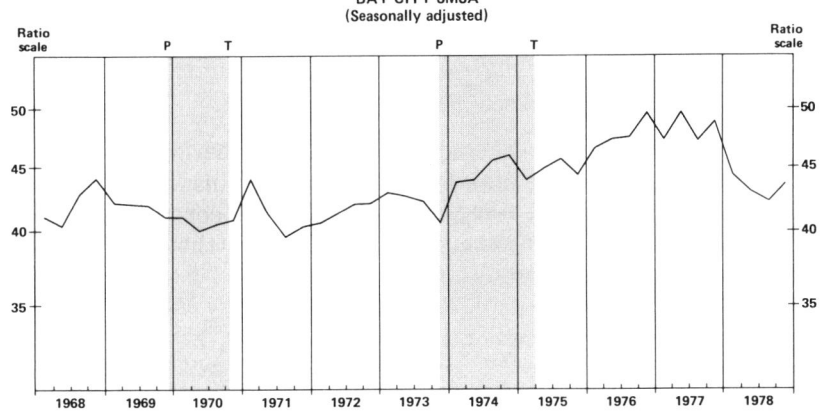

Source: Michigan Employment Security Commission.
Note: Seasonal adjustment by the W.E. Upjohn Institute.
Shaded areas indicate national recession periods as defined by the National Bureau of Economic Research, Inc.
P = peak and T = trough.

Construction

Construction activity in the Bay City SMSA exhibited a high degree of cyclical sensitivity, as measured by the behavior of new building permits for private housing and total employment in the local construction industry.

The Index of New Building Permits for private housing is shown in Chart BY5 for the 1965-77 period. Because of the volatility of the index from quarter to quarter (dashed line), a moving average was computed (solid line). The moving average reveals the short-run cyclical swings in the index. As is the case for other metropolitan areas in Michigan, the state as a whole, and the nation, new building permits began declining before peaks in national business activity, and the downswing during the late 1960s was less severe than that of the 1973-75 period. In the Bay City SMSA, the moving average of new building permits fell 54.4 percent during the former period and 72.3 percent during the latter. Both local contractions exceeded the downswings in the state and the nation.

The expansion in new building permits began in the second quarter of 1975, but fell during late 1976 and early 1977. Preliminary data for 1978 show that the moving average of new building permits has continued to fall. Despite this recent falloff, new building permits in the area grew at an annual rate of 3.2 percent over the 1965-1977 period, second only to the 8.4 percent growth rate for the Jackson SMSA.

Employment in the local construction industry is shown in Chart BY6. In this area, building trade employment exhibited a long cyclical downswing that began in the first quarter of 1971 and continued through the first quarter of 1976. That decline was 47.1 percent from peak to trough, second most severe among the 11 metropolitan areas in Michigan, and substantially larger than the statewide drop of 23.5 percent. Chart BY6 shows that construction employment experienced a strong recovery from 1976 to 1978, rising above its previous peak in 1971. For the nine-year period, construction employment in the Bay City SMSA grew at an annual rate of 1.8 percent, making this area one of only four SMSAs in Michigan where positive growth was evident over the 1970-78 period. Statewide, construction employment grew 0.4 percent per year, so that the Bay City SMSA outperformed the state as well.

Banking Activity

Indexes of two deposit series for commercial banks in the Bay

Bay City SMSA 73

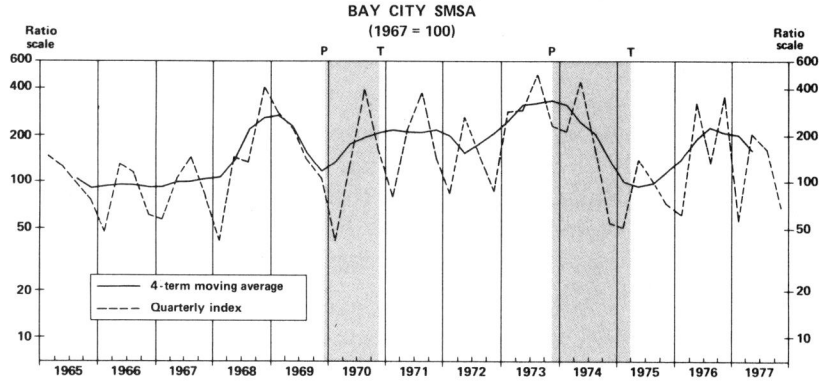

CHART BY 5
INDEX OF NEW BUILDING PERMITS, PRIVATE HOUSING
BAY CITY SMSA
(1967 = 100)

Source: The W.E. Upjohn Institute. Index based on the U.S. Department of Commerce, Bureau of the Census, *Construction Reports — Housing Authorized by Building Permits and Public Contracts*, C-40.

Shaded areas indicate national recession periods as defined by the National Bureau of Economic Research, Inc. P = peak and T = trough.

CHART BY 6
CONSTRUCTION EMPLOYMENT, BAY CITY SMSA
(Seasonally adjusted)

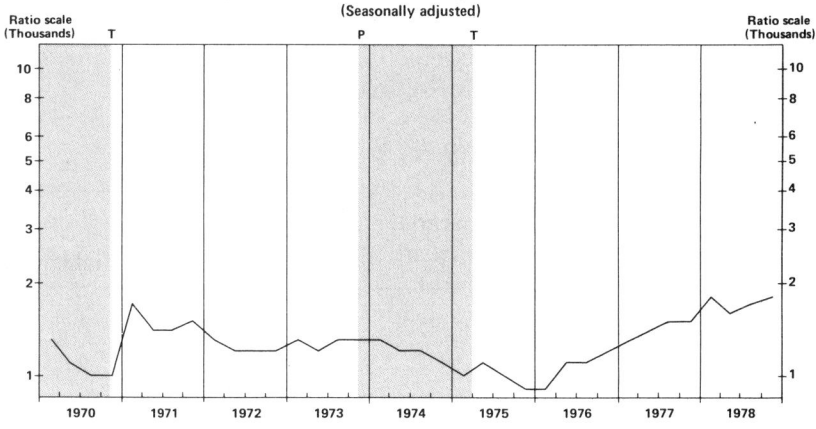

Source: Michigan Employment Security Commission.
Notes: Seasonal adjustment by the W.E. Upjohn Institute.

Shaded areas indicate national recession periods as defined by the National Bureau of Economic Research, Inc. P = peak and T = trough.

City SMSA are shown in Chart BY7. These represent the price-adjusted value of quarterly data for total deposits and demand deposits at the area's commercial banks expressed relative to the 1972 average. As the chart shows, both indexes began falling in late 1972, four quarters before the beginning of the national recession. The cyclical decline in the Index of Deflated Demand Deposits was more erratic and of longer duration than the slump in the Index of Deflated Total Deposits. From peak to trough, the latter fell 17.8 percent and the former 31.9 percent. Both contractions were the most severe among the state's 11 metropolitan areas.

As Chart BY7 shows, neither index has risen above pre-recession peaks. The Index of Deflated Total Deposits expanded 15.2 percent from its cyclical low to a level recorded in the second quarter of 1978. The expansion in the Index of Deflated Demand Deposits was somewhat less, 14.3 percent.

The large cyclical declines combined with the moderate expansion restricted growth of deposits. Table BY3 shows that, in current

TABLE BY3

Annual Rates of Growth of Selected Commercial Banking Indicators Bay City SMSA and Michigan[a]

(percent)

Indicator	Bay City SMSA	Michigan
Demand deposits (current dollars)	3.1	3.7
Deflated demand deposits[b]	-3.5	-3.0
Total deposits (current dollars)	7.0	7.3
Deflated total deposits[b]	0.1	0.4
Total loans (current dollars)	7.0	7.6
Commercial and industrial loans (current dollars)	3.3	7.3
Consumer installment loans (current dollars)	9.7	9.1

[a]Except where indicated otherwise, estimated growth rates are based on log-linear trends for the 1970-78 period.

[b]Current-dollar values adjusted for changes in the U.S. Consumer Price Index.

Bay City SMSA 75

CHART BY 7
INDEX OF DEFLATED TOTAL DEPOSITS AND INDEX OF DEFLATED DEMAND DEPOSITS
BAY CITY SMSA
(1972 = 100)
(Seasonally adjusted)

——— Total deposits index
- - - Demand deposits index

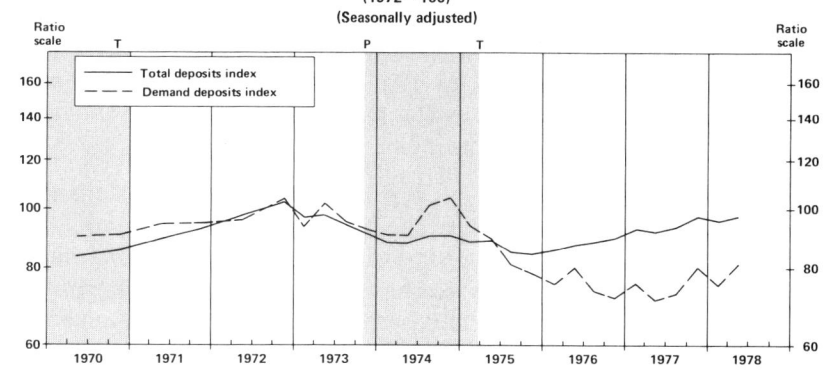

Source: The W.E. Upjohn Institute. Indices are based on data from the Federal Reserve Bank of Chicago.
Note: Shaded areas indicate national recession periods as defined by the National Bureau of Economic Research, Inc. P = peak and T = trough.

CHART BY 8
COMMERCIAL BANK LOANS, BAY CITY SMSA
(Current dollars)

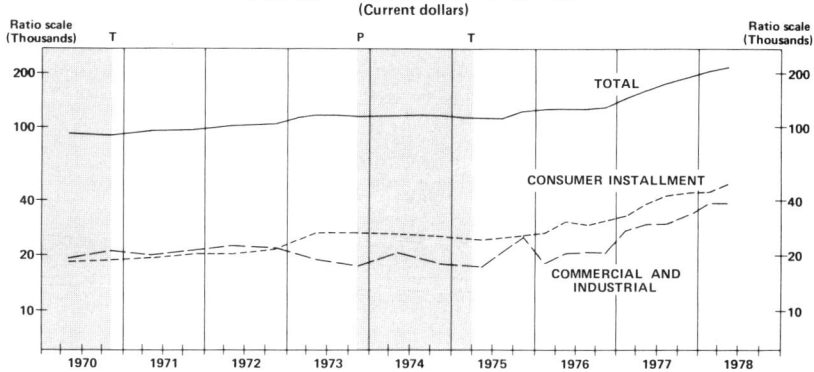

Source: Federal Reserve Bank of Chicago.
Notes: Seasonal adjustment of total loans by the W.E. Upjohn Institute. Other loans are not seasonally adjusted.
Shaded areas indicate national recession periods as defined by the National Bureau of Economic Research, Inc. P = peak and T = trough.

dollars, demand deposits and total deposits in the Bay City SMSA grew at an annual rate slightly below the statewide rate. After adjusting for price changes, however, demand deposits exhibited negative growth of 3.5 percent per year while total deposits displayed virtually no growth. The secular decline in deflated demand deposits in the Bay City SMSA exceeded the downtrends in all other metropolitan areas in the state.

The current-dollar value of total loans at commercial banks in the Bay City SMSA fell 4.1 percent during the 1973-75 recession, after rising about 26.2 percent during the preceding expansion period (see Chart BY8). The slump in commercial and industrial loans in the area was a major contributor to the falloff in total loans during the last recession. Measured from peak to trough, commercial and industrial loans fell 21.4 percent. That decline was second only to the 26.1 percent fall in such loans in the Battle Creek SMSA. As Chart BY8 shows, consumer installment loans suffered a milder contraction during the recession. However, the 9 percent decline in current-dollar volume was relatively severe among Michigan metropolitan areas, ranking fourth among the 11 SMSAs.

Chart BY8 shows that since hitting cyclical lows in 1975, the current-dollar volume of loans in the Bay City SMSA has expanded substantially. By the second quarter of 1978, commercial and industrial loans and consumer installment loans were up 117.6 percent and 100.8 percent, respectively. As a result, the current-dollar volume of total loans at commercial banks in the area rose 89 percent over a three-year period.

For the entire 1970-77 period, total loans grew at an annual rate of 7 percent, only slightly below the statewide growth rate of 7.6 percent (see Table BY3). Of the two major loan categories shown in Chart BY8, consumer installment loans exhibited the greater growth, increasing 9.7 percent per year compared to the more modest 3.3 percent pace for commercial and industrial loans.

Detroit SMSA

The Detroit SMSA, which is composed of six counties in southeastern Michigan, is by far the largest urban region in the state, with a population estimated at 4,370,200 in 1977.[1] In that year, 47.9 percent of Michigan's residents lived in the Detroit SMSA, and the area claimed 2 percent of the total U.S. population. The Detroit SMSA is one of the major metropolitan areas in the nation; in fact, in 1977 it ranked fifth in size among all SMSAs in the United States.

While this SMSA is the largest in Michigan, it is the only such area in the state to experience negative population growth during the 1970s. Between 1970 and 1977 the population is estimated to have fallen from 4,435,051 to 4,370,200, or 1.5 percent. That loss, attributed to a sizable net outmigration of 292,800 from the area, was due to the net loss of 371,500 in Wayne County, location of the City of Detroit, combined with a modest net inmigration of 78,700 into the other five more suburban counties. This intrametropolitan shift from central city to suburbs is a widely recognized urban phenomenon in the United States and is not unique to the Detroit SMSA. However, in the latter, the urban core in Wayne County contained 60.2 percent of the total SMSA population in 1970, and the 9.5 percent loss in that county—from 2,670,368 in 1970 to 2,417,700—more than offset gains in the other five counties over the seven-year period.

The relative size of the Detroit SMSA within Michigan is also reflected in the area's total personal income. In 1976, personal income amounted to $32.9 billion, which was 53.5 percent of Michigan's $61.5 billion total and more than nine times the personal income of the Flint SMSA, which ranked second. Also, the six-county Detroit SMSA accounted for 62.5 percent of the personal income generated in all metropolitan counties in the state in that year. However, the area's personal income growth between 1969 and 1976 ranked only ninth

---

1. The six counties that make up the Detroit SMSA are Wayne, Oakland, Livingston, Macomb, Lapeer, and St. Clair.

among the 11 SMSAs in Michigan, rising 7.8 percent per year in current dollars. Despite this relatively slow income growth, the Detroit SMSA has the highest per capita income among the state's metropolitan areas. In 1976 per capita income in the area was $7,496, exceeding the nationwide and statewide levels of $6,396 and $6,757, respectively.

The Detroit SMSA is a large, mature urban region. As such, the area experienced urban adjustment problems reflected, to some extent, in its intrametropolitan population shifts, overall population decline, and relatively slow income growth during the 1970s. Those phenomena tend to be related to long-run structural adjustments which the area has been undergoing for some time.[2]

The industrial structure of the Detroit SMSA is heavily concentrated in manufacturing. Table D1 shows that, in 1977, 33.7 percent of the area's total wage and salary workers were employed by manufacturing firms, primarily durable goods producers, compared to 32.4 percent for Michigan and almost 24 percent for the nation as a whole. The durable goods sector, which accounted

TABLE D1

Percentage Distribution of Total Wage and Salary Employment
Detroit SMSA and Michigan, 1972 and 1977

Item	Detroit SMSA 1972	Detroit SMSA 1977	Michigan 1972	Michigan 1977
Total Wage and Salary	100.0%	100.0%	100.0%	100.0%
Manufacturing	35.6	33.7	35.1	32.4
Durables	29.7	28.2	28.1	25.9
Nondurables	5.9	5.5	7.0	6.5
Nonmanufacturing	49.9	51.2	48.0	49.6
Government	14.5	15.0	16.9	18.0

Source: Michigan Employment Security Commission.

---

2. A brief description of the Detroit SMSA growth process appears in Shapiro, H.T., Mirowski, P.E., and Fulton, G.A., *Descriptive Study of the Major Labor Market Areas in Michigan,* Michigan Department of Labor, May 1978, pp. 4-35.

for about 84 percent of manufacturing employment in the Detroit SMSA, is dominated by automobile producers. In that year, the transportation equipment industry accounted for 41.9 percent of employment in the local manufacturing sector, compared to 34.7 percent statewide. Overall, the distribution of employment statewide is quite similar to that of the Detroit SMSA, although the latter has a smaller government sector and slightly larger manufacturing and nonmanufacturing sectors. However, the state's employment distribution is affected significantly by the Detroit SMSA due to its size. In 1977, the six-county area accounted for 51.4 percent of the state's manufacturing employment, 51.1 percent of nonmanufacturing employment, and 41.3 percent of government employment. Thus, the Detroit SMSAs absolute and relative size in terms of population, income, and employment makes it the dominant region in Michigan. In addition, its industrial structure, heavily concentrated in durable goods production, particularly automobiles, lends a great deal of cyclical instability to the local area and inevitably contributes significantly to the total economic performance in Michigan.

Business Conditions in the Detroit SMSA

Labor Market Conditions

The area's total wage and salary employment and its three major components—nonmanufacturing, manufacturing, and government employment—are shown by quarter in Chart D1. As is the case in the other metropolitan areas in Michigan, manufacturing employment exhibits the most cyclical sensitivity and largely accounted for the downswing in total wage and salary employment during the 1973-75 national recession. The cyclical contraction in local manufacturing employment coincided with the national business slump, dropping 20.1 percent during that period. While that decline exceeded the 18.7 percent drop statewide, it ranked only sixth in relative severity among the state's metropolitan areas. The Ann Arbor-Ypsilanti SMSA ranked first, with a decline in manufacturing employment of 26.0 percent, followed by Flint (25.5 percent), Jackson (25.5 percent), Lansing-East Lansing (24.5 percent), and Saginaw (20.6 percent). However, because of the size of the Detroit SMSA, job losses in the manufacturing sector amounted to approximately 125,000, or about 56 percent of total decline in manufacturing employment in Michigan during the recession.

80 Detroit SMSA

CHART D 1
WAGE AND SALARY EMPLOYMENT, DETROIT SMSA
(Seasonally adjusted)

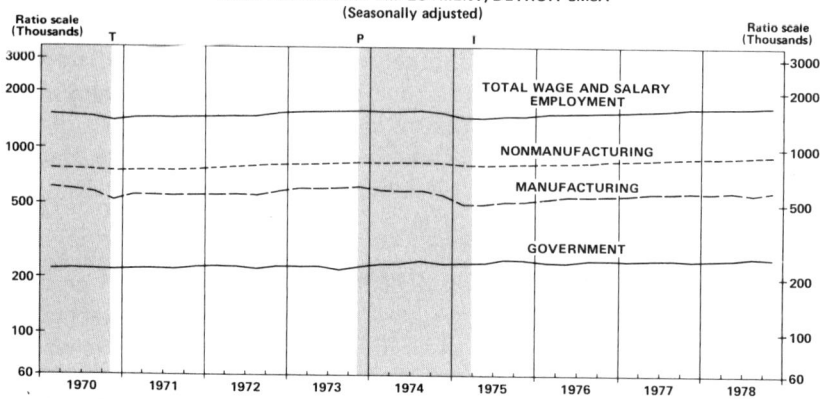

Source: Michigan Employment Security Commission.
Notes: Seasonal adjustment by the W.E. Upjohn Institute.
Shaded areas indicate national recession periods as defined by the National Bureau of Economic Research, Inc.
P = peak and T = trough.

CHART D 2
UNEMPLOYMENT RATE
DETROIT SMSA
(Seasonally adjusted)

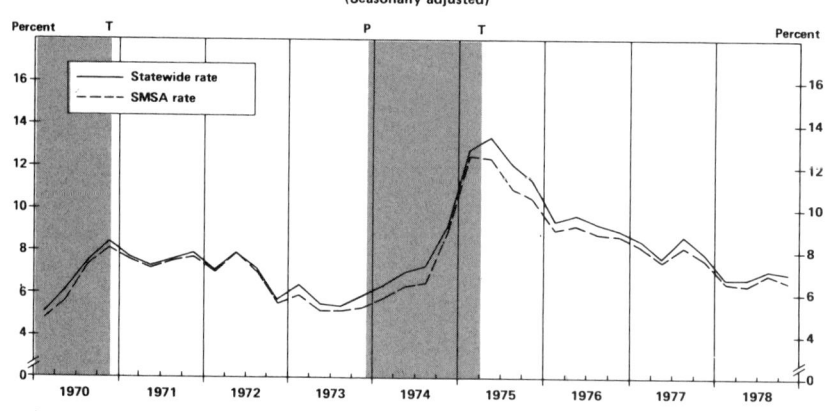

Source: Michigan Employment Security Commission.
Note: Seasonal adjustment by the W.E. Upjohn Institute.
Shaded areas indicate national recession periods as defined by the National Bureau of Economic Research, Inc.
P = peak and T = trough.

Nonmanufacturing employment in the Detroit SMSA fell 3.6 percent during the recession, but the downswing was brief, beginning in the third quarter of 1974 and ending in the second quarter of 1975. Chart D1 shows that government employment rose with little interruption throughout the recession and continued to expand through mid-1978.

The decline in total wage and salary employment in the area was 8.1 percent during the recession, the third most severe contraction among the state's 11 metropolitan areas and well above the 6.9 percent decline statewide. As was the case in all of the state's SMSAs, the relative stability of nonmanufacturing and government employment tempered the severity of the contraction in the manufacturing sector in the Detroit SMSA.

Following the last recession, total wage and salary employment has expanded 13.5 percent from the low in the first quarter of 1975 to the fourth quarter of 1978. Over the same period, the area's manufacturing employment increased 18.6 percent, but, as Chart D1 reveals, that still left employment in that sector below the pre-recession peak. Not so for nonmanufacturing employment, which expanded 13.2 percent to a level in the fourth quarter of 1978 well above its 1974 high. But, as is the case in other metropolitan areas, government employment rose the least after the recession, only 5.7 percent.

As Table D2 shows, employment growth in the Detroit SMSA during the 1970-78 period lagged behind the statewide growth rates. For the manufacturing sector, the slight downtrend of 0.1 percent over the nine-year period was the fourth poorest performance among the state's 11 SMSAs, ranking behind the negative growth in the Jackson SMSA (-1.5 percent), the Battle Creek SMSA (-1.4 percent), and the Muskegon SMSA (-1.1 percent).

Combined with only modest growth of jobs in the nonmanufacturing and government sectors, the manufacturing employment trend in this area resulted in an annual growth rate of just 1.3 percent for total wage and salary employment in the Detroit SMSA. Only the Battle Creek and Jackson SMSAs experienced slower growth during the period, 0.9 percent and 1.1 percent, respectively. Along with those two areas, the Detroit SMSA stands out as a relatively slow growth area in Michigan in terms of employment.

The unemployment situation in the Detroit SMSA is reflected in the behavior of two local indicators—the unemployment rate and

TABLE D2

Average Annual Growth Rates of Selected Labor Market Indicators Detroit SMSA and Michigan[a]

(percent)

Indicator	Detroit SMSA	Michigan
Total wage and salary employment	1.3	1.8
Manufacturing employment	-0.1	0.2
Nonmanufacturing employment	2.0	2.6
Government employment	2.0	3.0
Civilian labor force	1.3	2.0
Unemployment rate	2.6	2.8
Average weekly initial claims for UI[b]	6.7	7.3
Average workweek, production workers, mfg.[b]	0.4	0.6

[a] Except where indicated otherwise, estimated growth rates are based on log-linear trends for the 1970-78 period.

[b] Computed for the 1968-78 period.

initial claims for unemployment insurance. The area's jobless rate, shown in Chart D2, exhibited behavior quite similar to the state as a whole, beginning to rise before the national recession and increasing throughout the slump, then receding at about the time the recession was ending. This is exactly the same as the behavior of the unemployment rate nationwide. But, while the state and local jobless rates behaved similarly, the statewide average moved higher during the recession, reaching a seasonally adjusted 13.4 percent in the second quarter of 1975, compared to a cyclical high of 12.5 percent for the Detroit SMSA. Since the last recession, the area's jobless rate has fallen substantially, hitting a recent seasonally-adjusted low of 6.4 percent in the second quarter of 1978. Despite this decline, Chart D2 reveals an upward drift in the area's unemployment rate for the entire 1970-78 period. The uptrend was estimated at 2.6 percent per year, slightly below that of the state (see Table D2).

Average weekly initial claims for unemployment insurance,

shown in Chart D3, exhibit a similar pattern of leading behavior at peaks in national business activity and rough coincidence at troughs. Of course, because of the size of the Detroit SMSA, the movement of initial claims statewide is almost a mirror image of the behavior in the metropolitan area. During both recessions, however, the relative rise in average weekly initial claims was greater in the metropolitan area than in the state as a whole. Claims rose 237.3 percent and 375.1 percent in the Detroit SMSA during the 1969-70 and 1973-75 recessions, respectively, compared to 193.2 percent and 301.1 percent statewide. Despite the steeper rise locally, six other metropolitan areas exhibited greater increases than the Detroit SMSA during the 1969-70 recession, and five other areas had more severe upswings during the 1973-75 slump.[3]

Average weekly initial claims have fallen substantially since the last recession, as the local economy recovered and employment expanded. As of the fourth quarter of 1978, average weekly initial claims in the Detroit SMSA were 57.6 percent below their recession high. Statewide, initial claims were 53.7 percent below their cyclical high of the first quarter of 1975. However, as Chart D3 shows, at the end of 1978 initial claims locally and statewide were still well above their pre-recession levels. For the entire 1968-78 period, initial claims in the Detroit SMSA grew at an annual rate of 8.6 percent, which is above the state growth rate of 7.3 percent for that 11-year period. Compared to the other metropolitan areas in Michigan, the growth in initial claims in the Detroit SMSA ranked sixth, behind the uptrends in the Ann Arbor-Ypsilanti, Battle Creek, Jackson, Kalamazoo-Portage, and Muskegon SMSAs.

Finally, the average workweek of productions workers in the Detroit SMSA exhibited a high degree of cyclical sensitivity (see Chart D4). The average workweek began to decline before the start of the 1973-75 recession, falling from a seasonally-adjusted high of 45.7 hours in the first quarter of 1973 to 40 hours in the second quarter of 1975. That 12.5 percent drop in hours worked in local manufacturing industries ranked behind the declines in the Lansing-East Lansing, Flint, and Ann Arbor-Ypsilanti SMSAs. Each of those three areas is heavily concentrated in durable goods production, especially automobiles and related products, so that the relatively large falloff in hours worked during the last recession is not unexpected behavior. As Chart D4 shows, the average workweek has recovered substantially since the recession, and as of the last quarter of 1978 it was one hour below the pre-recession peak.

3. For the 1969-70 period those areas are Ann Arbor-Ypsilanti, Bay City, Flint, Jackson, Lansing-East Lansing, and Saginaw. During the 1973-75 recession, the increase in the Detroit SMSA was exceeded by upswings in Ann Arbor-Ypsilanti, Battle Creek, Flint, Grand Rapids, and Saginaw.

84 Detroit SMSA

CHART D 3
AVERAGE WEEKLY INITIAL CLAIMS FOR UNEMPLOYMENT INSURANCE, STATE PROGRAMS
DETROIT SMSA
(Seasonally adjusted)

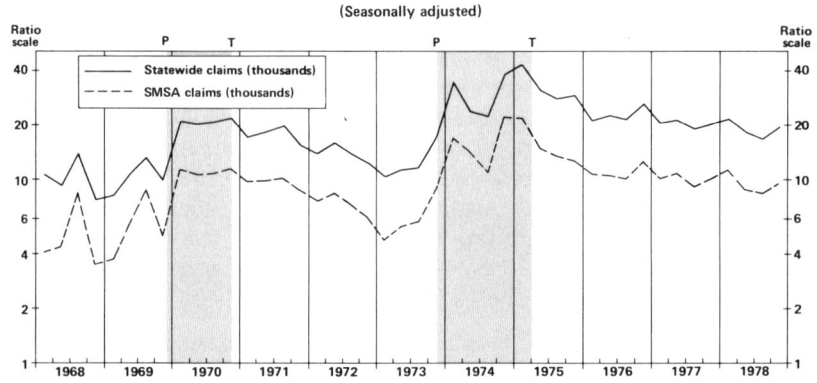

——— Statewide claims (thousands)
- - - - SMSA claims (thousands)

Source: The W.E. Upjohn Institute, based on data from the Michigan Employment Security Commission and the U.S. Department of Labor.
Shaded areas indicate national recession periods as defined by the National Bureau of Economic Research, Inc.
P = peak and T = trough.

CHART D 4
AVERAGE WEEKLY HOURS OF PRODUCTION WORKERS IN MANUFACTURING INDUSTRIES
DETROIT SMSA
(Seasonally adjusted)

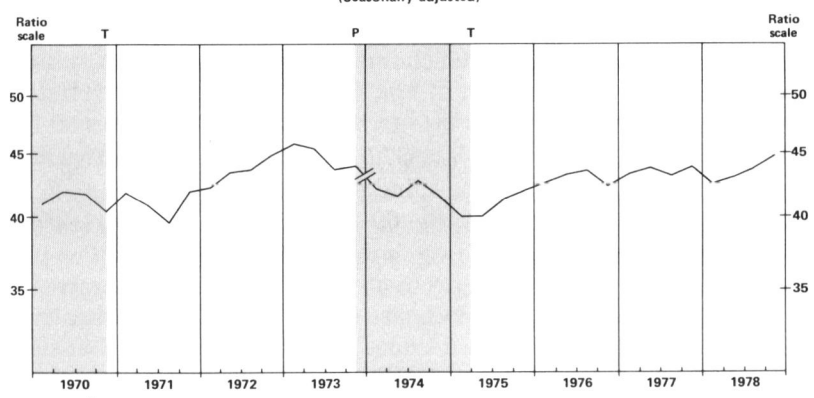

Source: Michigan Employment Security Commission.
Note: Seasonal adjustment by the W.E. Upjohn Institute.
Shaded areas indicate national recession periods as defined by the National Bureau of Economic Research, Inc.
P = peak and T = trough.
A slash mark indicates that data for earlier periods were not strictly comparable because of a change in the geographic area.

Construction

Construction activity in the Detroit SMSA is reflected in the behavior of new building permits for private housing and employment in the local construction industry. These two local indicators are shown in Charts D5 and D6.

The unadjusted quarterly values for new building permits are expressed relative to their 1967 average and displayed as the dashed line in Chart D5. That index exhibits a good deal of variability from quarter to quarter, but cyclical movements are evident. These are brought out more clearly by the moving average—the solid line in Chart D5. For the Detroit SMSA, the behavior of new building permits generally conformed to national business cycles, but with a lead, especially at peaks. A rather long lead occurred before the 1973 peak in national business activity. That is, new building permits in the six-county area began falling eight quarters, or two years, before the recession started. Nationwide, new building permits led by four quarters. The cyclical decline in the Detroit SMSA, therefore, began about one year before the contraction in a similar indicator for the nation as a whole. Within Michigan, the Detroit and Flint SMSAs exhibited the longest lead time for new building permits relative to the national recession.

However, in the Detroit SMSA the magnitude of the last cyclical decline was not large compared to most other metropolitan areas in Michigan. From its peak in the last quarter of 1971, the moving average shown in Chart D5 fell 67.5 percent to a low in the first quarter of 1975. While that decline exceeded the fall statewide of 56.2 percent, it was less than the 69.1 percent drop in the national index and ranks as the third mildest among the state's 11 metropolitan areas, behind the declines in similar indexes for the Jackson and Grand Rapids SMSAs.

Chart D5 shows that the new building permit index generally rose during the current expansion period. Preliminary data for 1978 indicate a slowing in the rate of advance of new building permits in the area and a leveling-off of the moving average. The apparent weakening in the upward momentum recently has left the Index of New Building Permits for the Detroit SMSA well below its pre-recession peak.

Employment in the construction industry in the Detroit SMSA, shown in Chart D6, experienced a fairly long decline during the mid-1970s. That slump began in early 1973 and ended in the second

Detroit SMSA

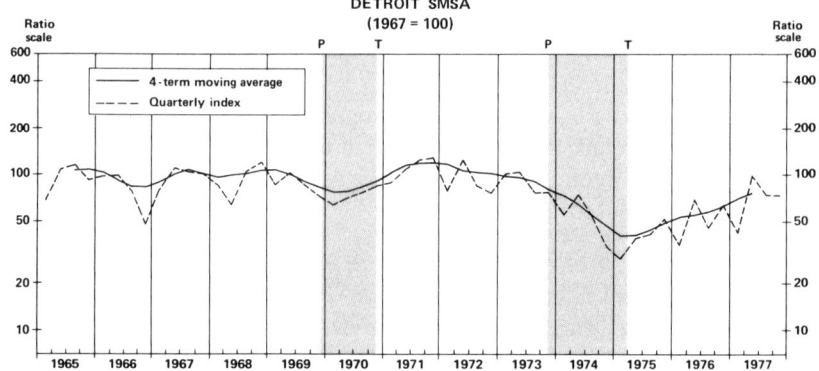

CHART D 5
INDEX OF NEW BUILDING PERMITS, PRIVATE HOUSING
DETROIT SMSA
(1967 = 100)

Source: The W.E. Upjohn Institute. Index based on the U.S. Department of Commerce, Bureau of the Census, *Construction Reports — Housing Authorized by Building Permits and Public Contracts, C-40.*

Shaded areas indicate national recession periods as defined by the National Bureau of Economic Research, Inc.
P = peak and T = trough.

CHART D 6
CONSTRUCTION EMPLOYMENT, DETROIT SMSA
(Seasonally adjusted)

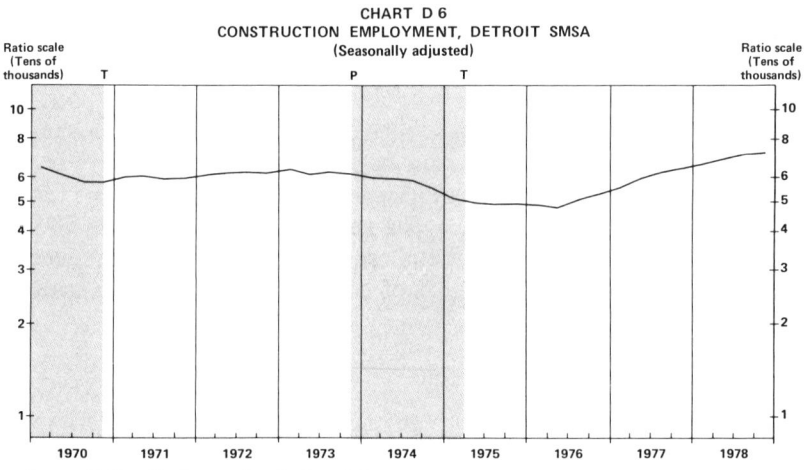

Source: Michigan Employment Security Commission.
Notes: Seasonal adjustment by the W.E. Upjohn Institute.
Shaded areas indicate national recession periods as defined by the National Bureau of Economic Research, Inc.
P = peak and T = trough.

quarter of 1976, lasting a total of 13 quarters. This long downswing was not unusual, since other Michigan metropolitan areas experienced contractions ranging from a high of 20 quarters in the Bay City SMSA to a relatively short seven quarters in the Muskegon-Norton Shores-Muskegon Heights SMSA. Among the 11 Michigan metropolitan areas, five experienced longer declines and five somewhat shorter contractions than the Detroit SMSA. Statewide, the cyclical decline in employment in the construction industry also lasted 13 quarters.

Measured from peak to trough, the magnitude of the decline in construction employment in the Detroit SMSA was 24.9 percent, only slightly above the 23.5 percent fall statewide. Six other metropolitan areas experienced more severe contractions. They are Ann Arbor-Ypsilanti, Battle Creek, Bay City, Flint, Jackson, and Saginaw.

As Chart D6 shows, construction employment in the area increased considerably since hitting a low in mid-1976. By the fourth quarter of 1978, local construction employment was up 52 percent above the recession low, and 14.1 percent above its pre-recession peak. Only two other SMSAs in the state had employment levels in 1978 above pre-recession peaks by more than the Detroit SMSA—Flint and Muskegon. For the 1970-78 period as a whole, the Detroit SMSA experienced no growth in construction employment, which, by the way, represents a fairly good performance over the nine-year period, because six metropolitan areas in the state experienced negative growth. The recovery in the Detroit SMSA since mid-1976 has offset the severe slump in construction employment and sizable falloff of construction activity resulting from the recession.

Banking Activity

The behavior of key banking indicators for the Detroit SMSA was not markedly different from the state as a whole during the 1970-78 period. This is not surprising because the size of the Detroit metropolitan area, with its large financial sector, suggests that activity there is likely to have a considerable influence on the state's aggregate performance. In 1977, for example, the Detroit SMSA accounted for 54 percent of the current-dollar volume of total deposits and 52 percent of total loans at commercial banks in the state. Table D3 shows that annual growth rates for deposits and loans at commercial banks over the eight-year period from 1970 to 1977 were slightly lower in the Detroit SMSA than in Michigan as a

88 Detroit SMSA

CHART D 7
INDEX OF DEFLATED TOTAL DEPOSITS AND INDEX OF DEFLATED DEMAND DEPOSITS
DETROIT SMSA
(1972 = 100)
(Seasonally adjusted)

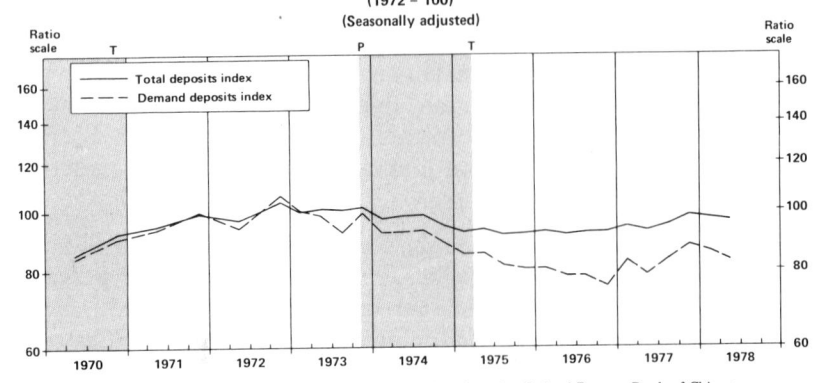

Source: The W.E. Upjohn Institute. Indices are based on data from the Federal Reserve Bank of Chicago.
Note: Shaded areas indicate national recession periods as defined by the National Bureau of Economic Research, Inc. P = peak and T = trough.

CHART D 8
COMMERCIAL BANK LOANS, DETROIT SMSA
(Current dollars)

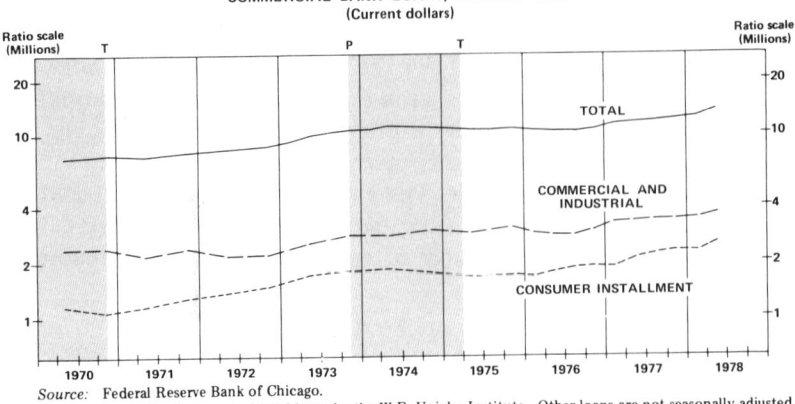

Source: Federal Reserve Bank of Chicago.
Notes: Seasonal adjustment of total loans by the W.E. Upjohn Institute. Other loans are not seasonally adjusted.
Shaded areas indicate national recession periods as defined by the National Bureau of Economic Research, Inc.
P = peak and T = trough.

TABLE D3

Average Annual Growth Rates of Selected Commercial Banking Indicators Detroit SMSA and Michigan[a]

(percent)

Indicator	Detroit SMSA	Michigan
Demand deposits (current dollars)	3.5	3.7
Deflated demand deposits[b]	-2.9	-3.0
Total deposits (current dollars)	6.5	7.3
Deflated total deposits[b]	-0.1	0.4
Total loans (current dollars)	6.3	7.6
Commercial and industrial loans (current dollars)	5.7	7.3
Consumer installment loans (current dollars)	8.7	9.1

[a] Except where indicated otherwise, estimated growth rates are based on log-linear trends for the 1970-78 period.

[b] Current-dollar values adjusted for changes in the Detroit Consumer Price Index for the Detroit SMSA and the U.S. Consumer Price Index for Michigan.

whole. For the banking indicators shown in Table D3, the largest difference between the local area and state was in the growth of loans; on a current-dollar basis loans grew at an annual rate of 7.6 percent statewide, compared to 6.3 percent in the Detroit SMSA. The latter was the slowest growth rate among the 10 metropolitan areas in the state for which comparable data were available.[4] Moreover, the growth of total deposits in the Detroit SMSA was the second lowest among these ten SMSAs. The Lansing-East Lansing SMSA had the lowest rate of growth—6.2 percent for the eight-year period. Therefore, the Detroit SMSA performed poorly relative to most other metropolitan areas in Michigan in terms of the growth of loans and deposits at commercial banks.

An Index of Deflated Demand Deposits and an Index of Deflated Total Deposits for the Detroit SMSA are shown in Chart D7.

---

4. The Battle Creek SMSA is excluded from the comparisons here because banking data for that area did not cover the entire 1970-77 period. (See the Battle Creek SMSA section.)

Quarterly values for both deposits series were deflated with the Consumer Price Index for Detroit, seasonally adjusted, and expressed relative to their 1972 average. As the chart reveals, the Index of Deflated Demand Deposits declined more than the Index of Deflated Total Deposits during the last recession. The former fell 29.0 percent and the latter a more moderate 11.8 percent. Besides the difference in the magnitude of the slump, the decline in the Index of Deflated Demand Deposits lasted 16 quarters (4 years!) compared to 11 quarters for the Index of Deflated Total Deposits. Except for the Bay City and Lansing-East Lansing SMSAs, those represent the longest contraction among metropolitan areas in Michigan. As noted in other sections, the lead at the 1973 peak in national business activity and the lag at the trough were characteristic of almost all local deposit series as well as the constant-dollar value of the national money supply, narrowly defined as cash outstanding plus demand deposits at commercial banks.

Chart D7 shows that neither index has moved above its pre-recession peak during the current expansion. The Index of Deflated Total Deposits hit a recent high of 99.1 in the fourth quarter of 1977, 2.5 percent below its previous peak. By comparison, the Index of Deflated Demand Deposits in the last quarter of 1977 was 17.1 percent below its pre-recession peak. In none of the 11 SMSAs in the state has the Index of Deflated Demand Deposits moved above pre-recession levels, so the behavior in the Detroit SMSA is not unusual.[5] However, in four areas, Ann Arbor-Ypsilanti, Flint, Grand Rapids, and Saginaw, the 1978 level of the Index of Deflated Total Deposits exceeds the pre-recession peak. Thus, while the Detroit behavior is not unique in this respect, it is certainly not among the stronger performances across the state.

The current-dollar volume of loans at commercial banks in the Detroit SMSA is shown by quarter in Chart D8, along with commerical and industrial loans and consumer installment loans. The latter two grew at annual rates of 5.7 percent and 8.7 percent, respectively, over the 1970-77 period. Commercial bank loans did suffer setbacks during the last recession but, as Chart D8 shows, the contractions in the area's loan volume began after the national recession was underway. This lagging tendency was evident at the trough also, with a recovery in the area's total loan volume not beginning until late 1976.

---

5. Sluggish growth of demand deposits is due to some extent to changes in financial laws and practices, as noted in Part I.

Measured from peak to trough, the current-dollar volume of loans fell 7 percent during the 1974-76 period, the second largest drop recorded among the state's metropolitan areas. The Battle Creek SMSA ranked first, with a decline of 9.8 percent. The declines in commercial and industrial loans and consumer installment loans were 7.3 percent and 9.8 percent, respectively.

As Chart D8 shows, loans increased considerably in the area in the last two and one-half years, and have moved well above their 1974 peak levels. In the second quarter of 1978 total loans were 20.2 percent above the cyclical peak in the second quarter of 1974. Commercial and industrial loans were 22.5 percent above their previous peak, and consumer installment loans 39.6 percent higher.

In sum, the Detroit SMSA experienced one of the larger declines in the current-dollar volume of loans at commercial banks during the 1973-75 recession. But, while the expansion has moved total loans above its previous peak, the area still ranks the lowest among the Michigan metropolitan areas in terms of growth during the 1970-77 period.

Flint SMSA

The Flint SMSA is made up of Shiawassee and Genesee counties and is the third largest metropolitan area in Michigan. In 1977, 514,400 people lived in the two-county area, up 1.1 percent from its 1970 level. In that same year, the Flint SMSA accounted for 5.6 percent of the total population in Michigan. The Flint SMSA is bordered by the Detroit SMSA to the southeast, the Saginaw SMSA to the north, and the Lansing-East Lansing SMSA to the west. It is, therefore, located in the heart of the automobile-producing region of the nation and is the center for the production of General Motors' cars. That one corporation employs about three-fourths of the manufacturing workers in the area.

The Flint SMSA ranks second among the state's 11 SMSAs in terms of total personal income, which was estimated at $3,627 million in 1976. In that year its per capita income was $7,046, ranking just behind Detroit ($7,496) and Ann Arbor-Ypsilanti ($7,215). Over the 1969-76 period, the area's personal income grew at an annual rate of 9.3 percent, which exceeds the 9.0 percent average growth rate for all U.S. SMSAs. Also, the personal income growth in the Flint SMSA was the fourth highest among the 11 Michigan SMSAs during that period. It should be noted, however, that because of its heavy dependence on automobile production, more than in any other Michigan metropolitan area, personal income in the Flint SMSA varies considerably over the business cycle. For example, in 1974, with the auto industry in the midst of a slump, personal income in the Flint SMSA rose only 1.7 percent from its 1973 level, the smallest income gain among the state's 11 metropolitan areas. But as automobile production boomed in 1976, personal income in the Flint SMSA rose 18.9 percent from its 1975 level, the largest annual increase among Michigan SMSAs.

94 Flint SMSA

The employment distribution in the Flint SMSA is heavily skewed toward manufacturing, automobiles in particular. Table F1 shows that in 1977, 43.6 percent of the area's total wage and salary employment was engaged in manufacturing, compared to 32.4 percent statewide. Both, of course, exceeded the nationwide proportion of 23.8 percent. Moreover, nearly all manufacturing employment in the Flint SMSA is accounted for by durable goods industries, dominated by automobile and related production at General Motors. In 1977, 72.8 percent of the manufacturing workers in the area were employed by the transportation equipment industry. Only the Lansing-East Lansing SMSA, with 63.8 percent of its manufacturing employment accounted for by the transportation equipment industry, approached anything like the Flint SMSA concentration of employment in a single industry.

The heavy concentration in automobile production in the Flint SMSA accounts for the area's relatively high personal income level because of the high wages and long workweek in the automobile industry. However, that industrial structure also imparts a high degree of cyclical volatility to the local economy, giving the Flint SMSA a reputation as a boom or bust area. The following sections reveal the cyclical sensitivity of economic activities in the area over roughly the last ten years.

TABLE F1

Percentage Distribution of Total Wage and Salary Employment Flint SMSA and Michigan, 1972 and 1977

Item	Flint SMSA 1972	Flint SMSA 1977	Michigan 1972	Michigan 1977
Total Wage and Salary	100.0%	100.0%	100.0%	100.0%
Manufacturing	45.0	43.6	35.1	32.4
Durables	42.4	41.4	28.1	25.9
Nondurables	2.6	2.2	7.0	6.5
Nonmanufacturing	41.2	41.9	48.0	49.6
Government	13.8	14.5	16.9	18.0

Source: Michigan Employment Security Commission.

Business Conditions in the Flint SMSA

Labor Market Conditions

Several indicators of employment and unemployment in the Flint SMSA reveal the high degree of cyclical sensitivity which characterizes this area's economy. Over roughly the last ten years the short-run behavior of local labor market indicators was substantially more volatile than those of the state as a whole. In fact, the Flint SMSA was the most cyclically sensitive of the state's 11 metropolitan areas; not an unexpected characteristic given the heavy concentration in automobile production within the area.

Chart F1 shows wage and salary employment in the Flint SMSA for the 1970-78 period. The cyclical phases of local total wage and salary and manufacturing employment conformed closely to national business cycle patterns, moving up during expansions and falling during recessions. In addition, the timing of peaks and troughs in the other two local employment series coincided with peaks and troughs in national business cycles over the period examined.

Not surprisingly, the behavior of total wage and salary employment in the Flint SMSA is largely determined by movements in local manufacturing employment. Note, for example, that the total wage and salary employment series shown in Chart F1 is nearly a mirror image of the manufacturing employment series. However, the latter is far more cyclically sensitive than the former. During the 1973-75 recession, for example, employment in local manufacturing industries fell 25.5 percent, the second most severe decline among the state's 11 metropolitan areas and ranking just behind the 26 percent decline in the Ann Arbor-Ypsilanti SMSA. On the other hand, nonmanufacturing employment fell only 5.5 percent during the recession, and government employment experienced a brief setback of 8.3 percent. Consequently, total wage and salary employment in the Flint SMSA declined by 13.4 percent, which is much less than that of manufacturing employment but still the most severe downswing among the state's 11 metropolitan areas. The slump in the Flint SMSA exceeded by a wide margin the 3.2 percent decline in the Kalamazoo-Portage SMSA, which recorded the mildest decline among the metropolitan areas in Michigan. Statewide, total wage and salary employment fell 6.9 percent during the 1973-75 period, just over one-half the magnitude of the downswing in the Flint SMSA.

96   Flint SMSA

CHART F 1
WAGE AND SALARY EMPLOYMENT, FLINT SMSA
(Seasonally adjusted)

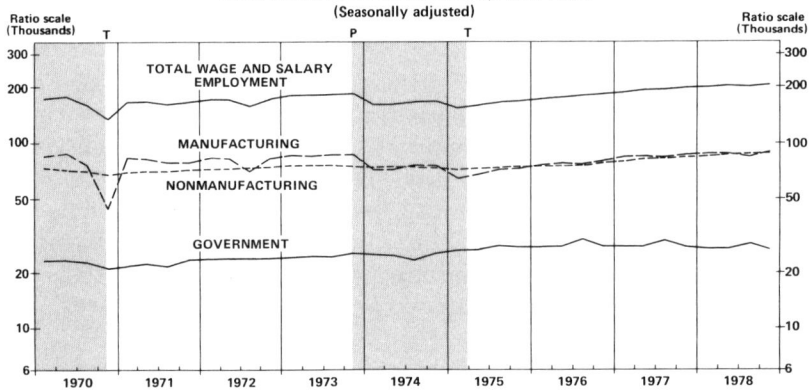

TOTAL WAGE AND SALARY EMPLOYMENT
MANUFACTURING
NONMANUFACTURING
GOVERNMENT

Source: Michigan Employment Security Commission.
Notes: Seasonal adjustment by the W.E. Upjohn Institute.
Shaded areas indicate national recession periods as defined by the National Bureau of Economic Research, Inc.
P = peak and T = trough.

CHART F 2
UNEMPLOYMENT RATE
FLINT SMSA
(Seasonally adjusted)

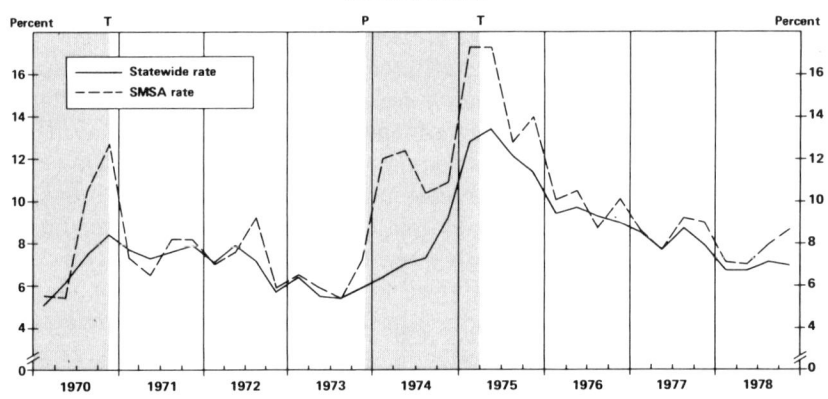

——— Statewide rate
--- SMSA rate

Source: Michigan Employment Security Commission.
Note: Seasonal adjustment by the W.E. Upjohn Institute.
Shaded areas indicate national recession periods as defined by the National Bureau of Economic Research, Inc.
P = peak and T = trough.

As Chart F1 shows, however, the employment expansions in the area following the national recessions have been fairly strong, with the major components of wage and salary employment moving above pre-recession peaks. During the recent expansion, manufacturing employment in the Flint SMSA rose 40.5 percent from its cyclical low in the first quarter of 1975 to the fourth quarter of 1978. That local expansion ranks just behind the 46.1 percent increase registered in the Ann Arbor-Ypsilanti SMSA. Over that time period, total wage and salary employment in the area increased 27.9 percent, the largest relative increase among the state's 11 metropolitan areas.

Despite the large cyclical swings that characterize employment in the Flint SMSA, the area's growth performance was relatively good. Table F2 shows that this metropolitan area outperformed the state in terms of growth of total wage and salary, manufacturing, and government employment over the 1970-78 period. The difference between the local area and the state was substantial for manufacturing employment, with the state moving upward at a slow 0.2 percent annual rate compared to 1.2 percent in the Flint SMSA. The latter ranks behind the 2.1 percent annual increase in the Ann Arbor-Ypsilanti SMSA, and the 1.8 percent annual increase in the Grand Rapids and Saginaw SMSAs. On the other hand, nonmanufacturing employment in the Flint SMSA grew at an annual rate of 2.4 percent, third lowest among the 11 metropolitan areas. Only the Battle Creek and Detroit SMSAs, with annual growth rates of 2.2 percent and 2.0 percent, respectively, exhibited slower growth of nonmanufacturing employment over the nine-year period.

Two indicators of the unemployment situation, the unemployment rate and average weekly initial claims for unemployment insurance, also reveal differences in growth and · cyclical performance between the state and the Flint SMSA. Table F2 shows that the jobless rate in the Flint SMSA drifted upward at an annual rate of 2.4 percent during the 1970-78 period, slightly below the 2.8 percent annual increase for the state as a whole. Both are below the uptrend of 4.3 percent per year that has characterized the national jobless rate. More striking, however, is the difference in the growth rates of initial claims for unemployment insurance. In the Flint SMSA such claims grew at an annual rate of 1.6 percent during the 1968-78 period, compared to the 7.3 percent rate for Michigan as a whole, and 7.1 percent nationwide. Moreover, the growth rate in the Flint SMSA was the lowest among the state's 11 metropolitan areas, far below the 11.1 percent annual growth rate in the Ann Arbor-Ypsilanti SMSA.

TABLE F2

Average Annual Growth Rates of
Selected Labor Market Indicators
Flint SMSA and Michigan[a]

(percent)

Indicator	Flint SMSA	Michigan
Total wage and salary employment	2.0	1.8
Manufacturing employment	1.2	0.2
Nonmanufacturing employment	2.4	2.6
Government employment	3.2	3.0
Civilian labor force.................	1.7	2.0
Unemployment rate.................	2.4	2.8
Average weekly initial claims for UI[b] ...	1.6	7.3
Average workweek, production workers, mfg.	1.5	0.6

[a] Except where indicated otherwise, estimated growth rates are based on log-linear trends for the 1970-78 period.

[b] Computed for the 1968-78 period.

However, both the unemployment rate and average weekly initial claims for unemployment insurance are more cyclically sensitive in the Flint SMSA than statewide. These two indicators are shown by quarter in Charts F2 and F3. Except for late 1970, the local and statewide jobless rates were quite similar between 1970 and 1973. But the national recession and the slump in the auto industry drove the jobless rate in the Flint SMSA from a seasonally adjusted 5.4 percent in the third quarter of 1973 to 17.3 percent in the second quarter of 1975. The latter was the highest jobless rate recorded among Michigan metropolitan areas and was well above the statewide high of 13.4 percent.

While the runup in the local jobless rate was considerable during the 1973-75 recession, Chart F2 shows that decline during the recent expansion was also of sizable magnitude. This coincides with

Flint SMSA 99

CHART F 3
AVERAGE WEEKLY INITIAL CLAIMS FOR UNEMPLOYMENT INSURANCE, STATE PROGRAMS
FLINT SMSA
(Seasonally adjusted)

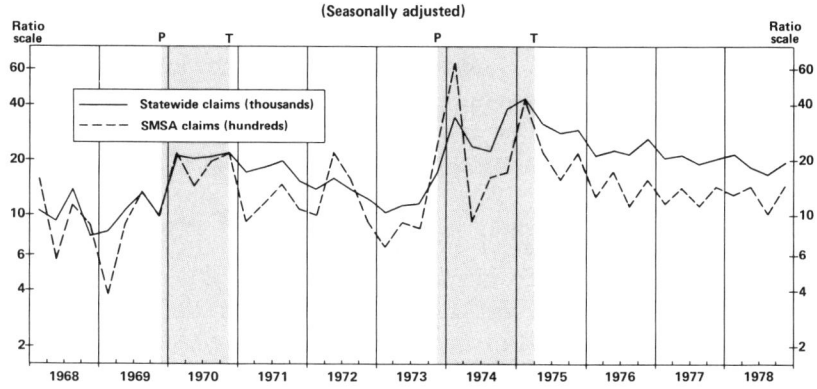

Source: The W.E. Upjohn Institute, based on data from the Michigan Employment Security Commission and the U.S. Department of Labor.

Shaded areas indicate national recession periods as defined by the National Bureau of Economic Research, Inc.
P = peak and T = trough.

CHART F 4
AVERAGE WEEKLY HOURS OF PRODUCTION WORKERS IN MANUFACTURING INDUSTRIES
FLINT SMSA
(Seasonally adjusted)

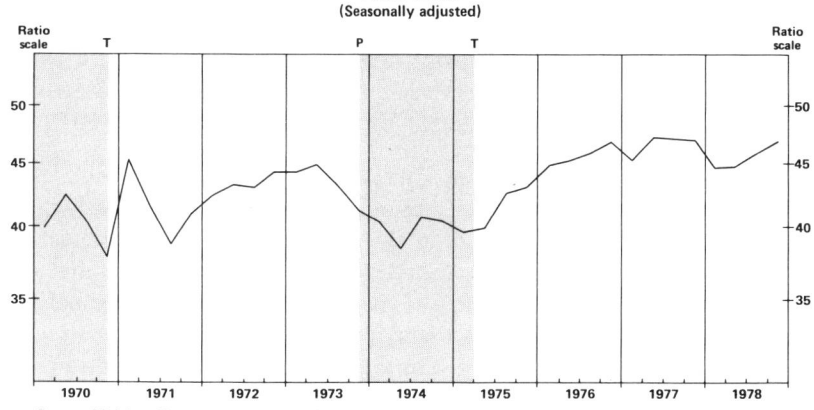

Source: Michigan Employment Security Commission.
Note: Seasonal adjustment by the W.E. Upjohn Institute.

Shaded areas indicate national recession periods as defined by the National Bureau of Economic Research, Inc.
P = peak and T = trough.

the relatively large cyclical upswings and downswings in employment in the area, and is further evidence of the overall cyclical sensitivity of this area. Despite the high unemployment rate which characterized this area in early 1975, the local jobless rate receded quickly to a level close to the statewide average by 1976. In that year the unemployment rate averaged 9.8 percent in the Flint SMSA compared to 9.4 percent for the state as a whole. Since then, the local jobless rate has remained just above the state average.

The short-run swings in average weekly initial claims for unemployment insurance for the state and the Flint SMSA are quite similar although, as Chart F3 reveals, the local series is more erratic. Both series lead at national business cycle peaks; that is, they begin to rise before a downswing in general business activity starts. Since peaks in total wage and salary and manufacturing employment tend to coincide with those of national business cycles, the upswing in average weekly initial claims tends to foreshadow slumps in employment in the Flint SMSA.[1] Finally, average weekly initial claims in the Flint SMSA fall substantially during expansion periods when employment in the area is rising. Chart F3 shows that while initial claims in the area moved downward after the 1969-70 and 1973-75 recessions, they failed to move below previous cyclical lows. By the end of 1978, initial claims in the Flint SMSA were down 63 percent from their first-quarter 1975 level. Statewide, initial claims had fallen 54 percent by the fourth quarter of 1978.

Finally, the seasonally-adjusted average workweek for manufacturing production workers, shown in Chart F4, experienced relatively large cyclical swings in the Flint SMSA, not unexpected behavior given the heavy concentration on automobile production in the area. The contraction in the average workweek began before the national recession started and prior to the beginning of the slump in local manufacturing employment. The decline was brief, however, lasting only four quarters. But its magnitude was large, as the workweek dropped from 44.9 hours in the second quarter of 1973 to 38.4 hours a year later. As was the case for other sensitive indicators in the metropolitan area, the average workweek recovered strongly, hitting a high of 46.9 hours in the fourth quarter

---

1. The erratic movements in the local series diminish, but do not eliminate, its usefulness as a leading indicator. Such behavior does indicate, however, that this series should be used cautiously and certainly in combination with other sensitive indicators, the average workweek, for example.

of 1978. That represents the longest workweek among manufacturing workers in any of the state's metropolitan areas, and is well above the state average of 43.6 hours for the last three months of 1978.

Construction

A quarterly Index of New Building Permits for private housing is shown in Chart F5 (dashed line). Each quarterly value is expressed in terms of the 1967 average so that the base value of that year equals 100. The solid line in the chart is a moving average of the unadjusted values and is much smoother than the unadjusted index. Chart F5 reveals that new building permits in the Flint SMSA did not decline during the 1969-70 recession. In fact, the moving average rose fairly steadily from the last quarter of 1966 to the fourth quarter of 1971. The Ann Arbor-Ypsilanti SMSA was the only other metropolitan area in Michigan that did not experience a recession-related contraction in new building permits in 1969-70. Statewide, new building permits fell 21.5 percent and the national Index of New Building Permits dropped 24.5 percent. Thus, local commitments to build private housing, reflected in the Index of New Building Permits, were unaffected by the relatively mild national recession in 1969-70.

However, the more recent slump in building permit activity was not only severe, but was also quite long. Chart F5 shows that the moving average fell from a peak in the last quarter of 1971 to a low in the first quarter of 1975. That local cyclical decline, which lasted more than three years, equaled the duration of the slump in the adjacent Detroit and Saginaw SMSAs and was the longest among the state's metropolitan areas. The magnitude of the decline in the Flint SMSA was 79.6 percent measured from peak to trough, and that was exceeded only by the fall of 88.6 percent and 84.5 percent in the Ann Arbor-Ypsilanti and Battle Creek SMSAs, respectively.

Chart F5 shows, however, that the subsequent expansion in new building permits has been fairly strong. Preliminary data for 1978 show that the moving average of new building permits has continued to increase, with its latest level being 195 percent above the recession low. Despite this expansion, new building permits in the Flint SMSA have failed to reach their pre-recession high, a situation that existed in all Michigan metropolitan areas at the end of 1978.

Employment in the local construction industry is shown in Chart F6. As is the case in other metropolitan areas, construction employ-

Flint SMSA

CHART F 5
INDEX OF NEW BUILDING PERMITS, PRIVATE HOUSING
FLINT SMSA
(1967 = 100)

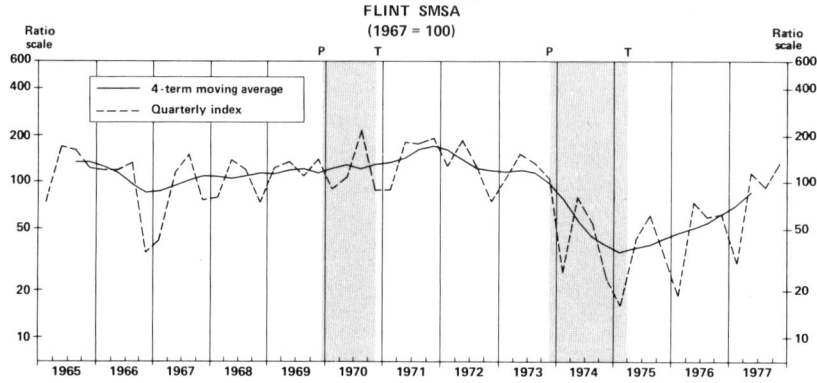

Source: The W.E. Upjohn Institute. Index based on the U.S. Department of Commerce, Bureau of the Census, *Construction Reports — Housing Authorized by Building Permits and Public Contracts, C-40.*
Shaded areas indicate national recession periods as defined by the National Bureau of Economic Research, Inc.
P = peak and T = trough.

CHART F 6
CONSTRUCTION EMPLOYMENT, FLINT SMSA
(Seasonally adjusted)

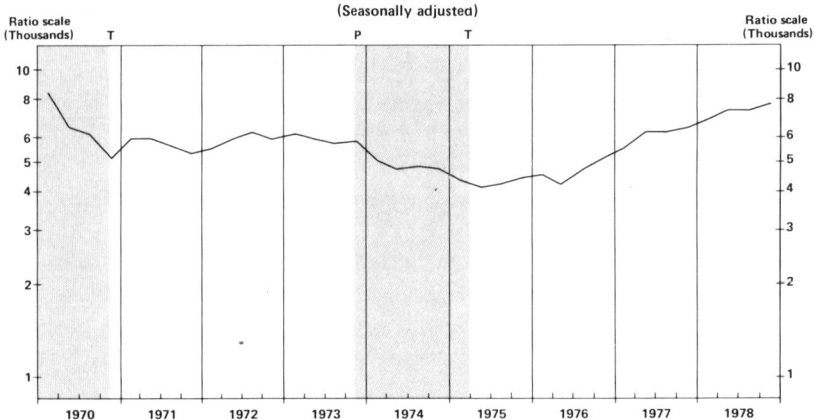

Source: Michigan Employment Security Commission.
Notes: Seasonal adjustment by the W.E. Upjohn Institute.
Shaded areas indicate national recession periods as defined by the National Bureau of Economic Research, Inc.
P = peak and T = trough.

ment in the Flint SMSA exhibits a high degree of cyclical sensitivity. After a modest expansion following the 1969-70 recession, construction employment began declining in the third quarter of 1972—five quarters before the national recession started—and continued to do so until the second quarter of 1975. Over that period employment in the construction industry fell 33.9 percent, exceeding the statewide drop of 23.5 percent. However, three other metropolitan areas in the state suffered more severe contractions than Flint. They are Ann Arbor-Ypsilanti, Jackson, and Saginaw.

As Chart F6 shows, the recent expansion in employment in the local construction industry has been quite strong. Employment rose 87.8 percent from the second quarter of 1975 to the fourth quarter of 1978. That is the most vigorous expansion for this series among the state's 11 metropolitan areas. As a result, construction employment in the Flint SMSA was well above its pre-recession peak by the fourth quarter of 1978. Only in the Muskegon SMSA has construction employment moved further above its pre-recession high.

Banking Activity

The behavior of total deposits and demand deposits at commercial banks in the Flint SMSA is revealed in Chart F7. Each series has been adjusted for price changes and expressed relative to its 1972 average.

The cyclical declines in both indexes began before the 1973 peak in national business activity. However, the downswing in the Index of Deflated Demand Deposits started two quarters before that of the Index of Deflated Total Deposits, and the former was more precipitous than the latter. From peak to trough, the Index of Deflated Demand Deposits fell 26.7 percent compared to the 13.4 percent decline in the Index of Deflated Total Deposits. Neither of these two contractions was the most severe among the 11 Michigan metropolitan areas.

As was the case for several local employment and construction indicators, the expansion of commercial bank deposits since the last recession has been relatively strong in the Flint SMSA. In the second quarter of 1978, the latest quarterly data available, the Index of Deflated Demand Deposits was 26.7 percent above its recession low, and the Index of Deflated Total Deposits had risen 18.7 percent. And, as Chart F7 shows, the latter had risen above its

104 Flint SMSA

CHART F 7
INDEX OF DEFLATED TOTAL DEPOSITS AND INDEX OF DEFLATED DEMAND DEPOSITS
FLINT SMSA
(1972 = 100)
(Seasonally adjusted)

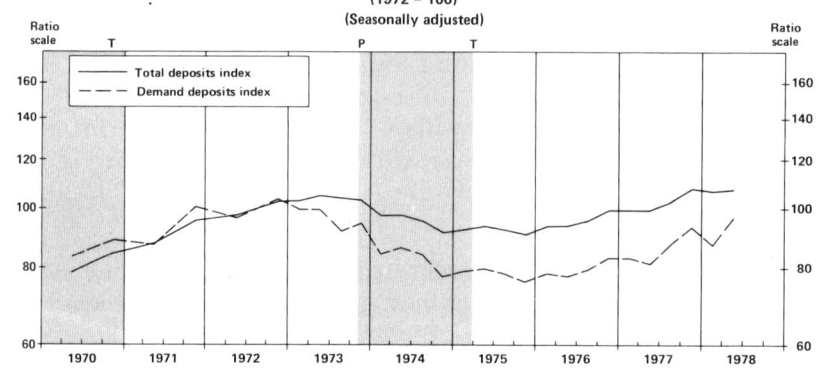

Source: The W.E. Upjohn Institute. Indices are based on data from the Federal Reserve Bank of Chicago.
Note: Shaded areas indicate national recession periods as defined by the National Bureau of Economic Research, Inc. P = peak and T = trough.

CHART F 8
COMMERCIAL BANK LOANS, FLINT SMSA
(Current dollars)

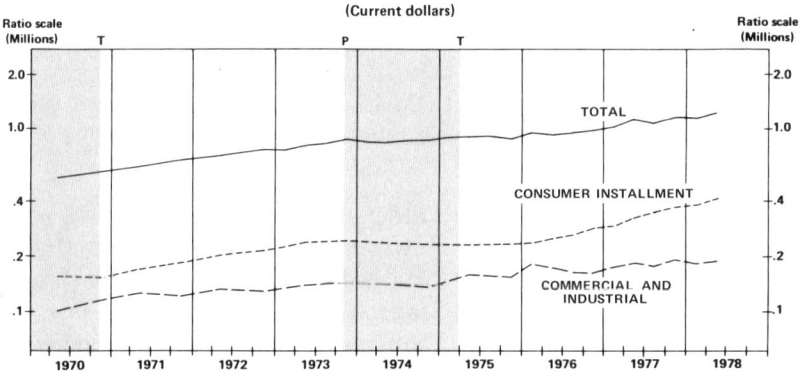

Source: Federal Reserve Bank of Chicago.
Notes: Seasonal adjustment of total loans by the W.E. Upjohn Institute. Other loans are not seasonally adjusted.
Shaded areas indicate national recession periods as defined by the National Bureau of Economic Research, Inc.
P = peak and T = trough.

pre-recession peak by the end of 1977, joining the Ann Arbor-Ypsilanti, Grand Rapids, and Saginaw SMSAs as an area in which the deflated value of total deposits at commercial banks had moved above its previous cyclical high. Although the expansion in this area's demand deposits index was the most vigorous among Michigan metropolitan areas, in mid-1978 the index was still 7.1 percent below its pre-recession peak.

Table F3 shows that the annual growth rate of the deposit series in the Flint SMSA exceeded the growth rate for the state as a whole. But the growth of total deposits in the Flint SMSA, 9 percent in current dollars and 2 percent in constant dollars, ranked only fifth among Michigan metropolitan areas.

TABLE F3

Average Annual Growth Rates of Selected
Commercial Banking Indicators
Flint SMSA and Michigan[a]

(percent)

Indicator	Flint SMSA	Michigan
Demand deposits (current dollars)	4.6	3.7
Deflated demand deposits[b]	-2.2	-3.0
Total deposits (current dollars)	9.0	7.3
Deflated total deposits[b].............	2.0	0.4
Total loans (current dollars)	8.7	7.6
Commercial and industrial loans (current dollars)	7.1	7.3
Consumer installment loans (current dollars)	9.3	9.1

[a] Except where indicated otherwise, estimated growth rates are based on log-linear trends for the 1970-78 period.

[b] Current-dollar values adjusted for changes in the U.S. Consumer Price Index.

The growth of loans was also relatively slow in the Flint SMSA, increasing in current-dollar terms at an annual rate of 8.7 percent over the 1970-77 period. While that exceeded the statewide growth rate of 7.6 percent, it ranked only fifth behind the growth in Saginaw, Muskegon, Ann Arbor-Ypsilanti, and Kalamazoo-Portage. For those same years, the current-dollar volume of commercial and industrial loans grew at an annual rate of 7.1 percent. In this case, only the Bay City, Detroit, and Jackson SMSAs registered slower growth. In addition, the 9.3 percent annual growth rate of consumer installment loans ranked eighth among the 10 metropolitan areas in the state for which comparable data were available. Note that in terms of the latter two loan series the Flint performance was not much different from the state.

Chart F8 shows that the growth rate of total loans slowed considerably during the 1973-75 recession and that consumer installment loans in the area experienced an absolute decline throughout that period. That contraction was a relatively mild 4.8 percent, however. In current dollars, consumer installment loans dropped 6 percent. Only in the Ann Arbor-Ypsilanti SMSA, where such loans fell 4.3 percent, and in the Kalamazoo-Portage SMSA, where almost no decline was evident, did consumer installment loans experience milder setbacks during the last recession. The recession performance of commercial bank loans in the Flint SMSA is remarkable, given the relatively large contraction in the area's employment during that business slump.

Also, Chart F8 reveals that since the end of the recession the expansion of loans has been strong. Commercial and industrial loans are well above their 1974 levels, and consumer installment loans have risen 81 percent since hitting a recession low in the second quarter of 1975. And in the second quarter of 1978, total loans at commercial banks in the Flint SMSA were up 35.7 percent over their pre-recession level.

In summary, the Flint SMSA exhibited a high degree of cyclical sensitivity during the 1970s, characterized by sizable downswings and upswings in key economic indicators. The latter typically offset the previous cyclical decline, sending local economic activities above their previous peaks. This relatively large cyclical amplitude was most evident in the labor market indicators. Overall, the Flint SMSA can easily be classified as the most cyclically sensitive metropolitan area in Michigan in terms of employment. However, while deposits and loans at commercial banks in the area did experience cyclical swings, the Flint SMSA was not the most volatile metropolitan area in this regard during the 1970-78 period, nor was it subject to the largest fluctuations in construction activity.

Grand Rapids SMSA

In terms of population, the Grand Rapids SMSA is the second largest metropolitan area in Michigan. In 1977, 575,600 people lived in Ottawa and Kent counties, which make up the Grand Rapids SMSA. That represents 6.3 percent of the state's total population in that year. Although this area ranked second to the Detroit SMSA in size, it was considerably smaller in absolute terms, since the latter had a population of 4,370,200 in 1977, more than seven times that of the Grand Rapids SMSA.

Between 1970 and 1977, population increased 6.7 percent in this two-county area. That gain ranked third among Michigan metropolitan areas, just behind the increase of 7.3 percent and 6.9 percent in the Lansing-East Lansing and Ann Arbor-Ypsilanti SMSAs, respectively, and well above the statewide increase of 2.8 percent for the eight-year period.

The Grand Rapids SMSA also ranked third among the state's metropolitan areas in terms of total personal income, which amounted to $3,553 million in 1976. In that year, the Grand Rapids SMSA accounted for 5.8 percent of total personal income in Michigan. However, this metropolitan area had the third lowest per capita income among the 11 metropolitan areas. In 1976, its per capita income of $6,254 exceeded that of the Muskegon-Norton Shores-Muskegon Heights SMSA ($5,496) and the Bay City SMSA ($6,116), and was below the statewide level of $6,757. The relatively low per capita income level in the Grand Rapids SMSA is attributable largely to the fact that it is not a high-wage area. This, of course, is due to the local industry mix, which is fairly diversified and not heavily dependent on the high-wage automobile industry. In fact, the 1976 average wage rate in this area's manufacturing sector was the lowest among the state's metropolitan areas.

Table GR1 shows that the employment distribution in the local area is similar to the state, except that the government sector is substantially smaller in the Grand Rapids SMSA. In 1977, the proportion of total wage and salary employment accounted for by durable goods industries was nearly the same in the local area and the state as a whole, 26.1 percent and 25.9 percent, respectively. However, the aggregates shown in Table GR1 mask the major difference between this area and the state: the key durable goods industries in the Grand Rapids SMSA are not linked to automobile production. In 1977, the transportation equipment industry accounted for only 5 percent of manufacturing employment in the area, compared to 34.7 percent statewide and 72.8 percent in the Flint SMSA. The fabricated metals industry accounted for 19.8 percent of the area's manufacturing employment in 1977, followed by furniture and fixtures, 14.7 percent, and nonelectrical machinery, 14.0 percent. More than one-half of the jobs in the Michigan furniture industry are located in the Grand Rapids SMSA. Thus, the local area has an industry mix substantially different from the state as a whole or from most other metropolitan areas in Michigan. This is especially the case when the Grand Rapids SMSA is compared to the metropolitan areas in the eastern part of the state. In addition, the manufacturing sector in the Grand Rapids SMSA is somewhat more diversified, with no single industry or firm accounting for the bulk of employment in the area. Only the Kalamazoo-Portage SMSA has a more diversified industrial base than that of the Grand Rapids SMSA.

TABLE GR1

Percentage Distribution of Total Wage and Salary Employment
Grand Rapids SMSA and Michigan, 1972 and 1977

Item	Grand Rapids SMSA 1972	Grand Rapids SMSA 1977	Michigan 1972	Michigan 1977
Total Wage and Salary	100.0%	100.0%	100.0%	100.0%
Manufacturing	37.5	34.6	35.1	32.4
Durables	28.2	26.1	28.1	25.9
Nondurables	9.3	8.5	7.0	6.5
Nonmanufacturing	51.9	52.9	48.0	49.6
Government	10.6	12.5	16.9	18.0

Source: Michigan Employment Security Commission.

Business Conditions in the Grand Rapids SMSA

Labor Market Conditions

In terms of growth and cyclical stability, the Grand Rapids SMSA outperformed all other Michigan metropolitan areas. Table GR2 shows that over the 1970-78 period, total wage and salary employment grew at an annual rate of 3.1 percent, which is considerably above the statewide rate of 1.8 percent and represents the most rapid growth among the state's 11 metropolitan areas over that nine-year period. The 1.8 percent growth rate of manufacturing employment in the Grand Rapids SMSA ranks second, just behind the 2.1 percent growth rate in the Ann Arbor-Ypsilanti SMSA. However, while manufacturing and nonmanufacturing employment in the area grew at relatively rapid rates, the largest contributor to local employment growth was the government sector. For the nine-year period beginning in 1970, government employment grew at an annual rate of 6.1 percent, which is the highest growth rate for public employment among the state's metropolitan areas.

Table GR2 shows that the Grand Rapids SMSA also outperformed the state in terms of growth of the civilian labor force,

TABLE GR2

Average Annual Growth Rates of Selected Labor Market Indicators Grand Rapids SMSA and Michigan[a]

(percent)

Indicator	Grand Rapids SMSA	Michigan
Total wage and salary employment	3.1	1.8
Manufacturing employment	1.8	0.2
Nonmanufacturing employment	3.4	2.6
Government employment	6.1	3.0
Civilian labor force	2.9	2.0
Unemployment rate	−0.6	2.8
Average weekly initial claims for UI[b]	3.7	7.3
Average workweek, production workers, mfg.	0.3	0.6

[a]Except where indicated otherwise, estimated growth rates are based on log-linear trends for the 1970-78 period.

[b]Computed for the 1968-78 period.

the unemployment rate, and average weekly initial claims for unemployment insurance. Over the 1970-78 period the jobless rate in the area exhibited a slight downtrend of 0.6 percent per year, compared to the upward drift of 2.8 percent for the statewide average. Thus, the Grand Rapids SMSA joins the Bay City SMSA as the only metropolitan area in Michigan not exhibiting positive growth in the local unemployment rate. In line with the area's relatively strong employment growth, initial claims for unemployment insurance grew at a modest rate of 3.7 percent per year over the 1968-78 period. Only the Flint SMSA had a lower growth rate, 1.6 percent. For Michigan as a whole, initial claims expanded 7.3 percent per year over that eleven-year period.

Chart GR1 shows that nonmanufacturing and government employment were highly stable throughout the 1970-78 period, although the former experienced a slight 1.2 percent decline during the 1973-75 recession. But that decline was the mildest among the state's metropolitan areas except for the Kalamazoo-Portage SMSA, in which nonmanufacturing employment suffered no setback during the recession. On the other hand, manufacturing employment fell 18.3 percent in the Grand Rapids SMSA during the last recession, and that largely accounted for the 5.1 percent decline in the area's total wage and salary employment. Both local contractions were less than the statewide declines of 18.7 percent and 6.9 percent for manufacturing and total wage and salary employment, respectively.

Since the recession, employment in each sector has increased substantially. As of the fourth quarter of 1978, manufacturing employment was up 29 percent over its recession low and had reached a level that was 5.4 percent above its pre-recession peak. Combined with the steady growth in nonmanufacturing and government employment, the expansion in manufacturing has resulted in a 17.5 percent increase in the area's total wage and salary employment since the last recession ended.

The jobless rate in the Grand Rapids SMSA exhibits about the same cyclical sensitivity as the statewide rate but, as Chart GR2 shows, the local rate has remained somewhat below the state average throughout the nine-year period. During the last recession the local unemployment rate rose from a seasonally-adjusted 5 percent in the third quarter of 1973 to 12.4 percent in the second quarter of 1975. Since that time, the unemployment rate in the Grand Rapids SMSA has fallen considerably, hitting a recent low of 4.8 percent in the first quarter of 1978, below its pre-recession low.

Grand Rapids SMSA

CHART GR 1
WAGE AND SALARY EMPLOYMENT, GRAND RAPIDS SMSA
(Seasonally adjusted)

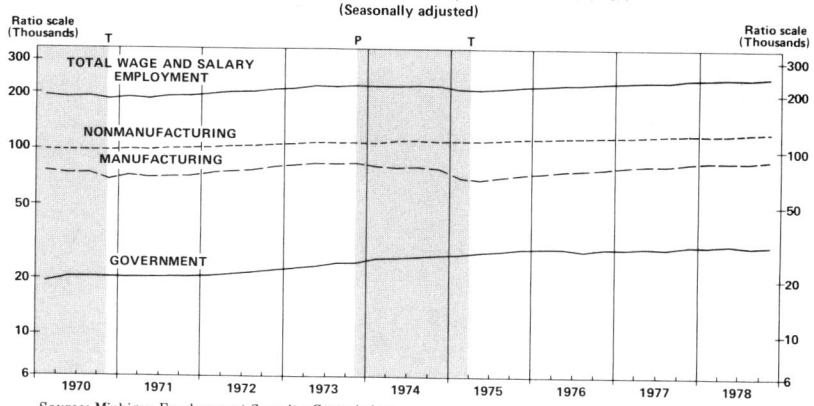

Source: Michigan Employment Security Commission.
Notes: Seasonal adjustment by the W.E. Upjohn Institute.
Shaded areas indicate national recession periods as defined by the National Bureau of Economic Research, Inc.
P = peak and T = trough.

CHART G 2
UNEMPLOYMENT RATE
GRAND RAPIDS SMSA
(Seasonally adjusted)

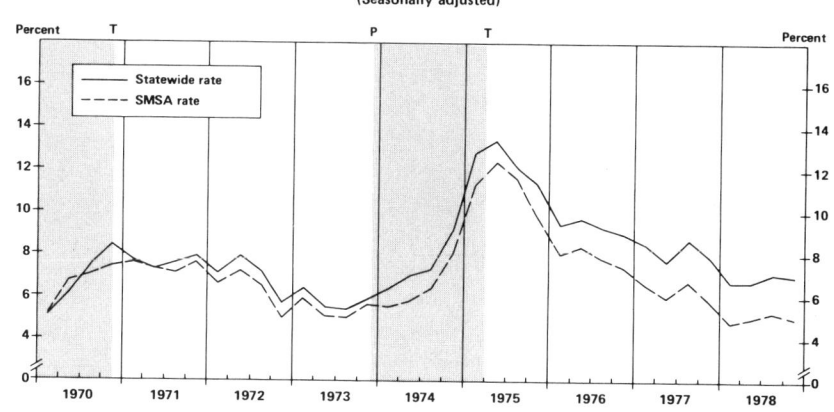

Source: Michigan Employment Security Commission.
Note: Seasonal adjustment by the W.E. Upjohn Institute.
Shaded areas indicate national recession periods as defined by the National Bureau of Economic Research, Inc.
P = peak and T = trough.

Grand Rapids SMSA

CHART G 3
AVERAGE WEEKLY INITIAL CLAIMS FOR UNEMPLOYMENT INSURANCE, STATE PROGRAMS
GRAND RAPIDS SMSA
(Seasonally adjusted)

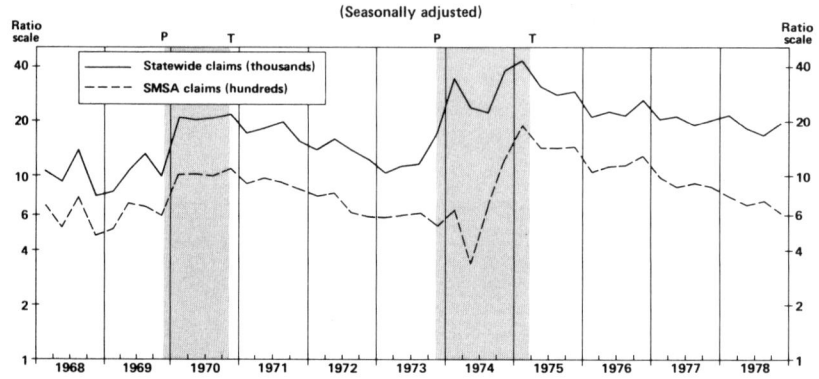

Source: The W.E. Upjohn Institute, based on data from the Michigan Employment Security Commission and the U.S. Department of Labor.
Shaded areas indicate national recession periods as defined by the National Bureau of Economic Research, Inc.
P = peak and T = trough.

CHART G 4
AVERAGE WEEKLY HOURS OF PRODUCTION WORKERS IN MANUFACTURING INDUSTRIES
GRAND RAPIDS SMSA
(Seasonally adjusted)

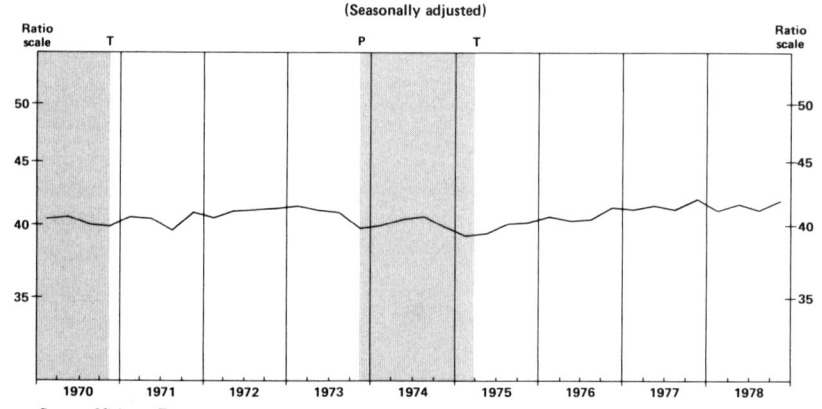

Source: Michigan Employment Security Commission.
Note: Seasonal adjustment by the W.E. Upjohn Institute.
Shaded areas indicate national recession periods as defined by the National Bureau of Economic Research, Inc.
P = peak and T = trough.

Grand Rapids is, therefore, the only metropolitan area in Michigan where the jobless rate has fallen below its pre-recession level.

Average weekly initial claims for unemployment insurance, shown in Chart GR3, exhibit an overall pattern of behavior in the Grand Rapids SMSA that is also quite similar to the state as a whole. The local series was more cyclically volatile, however. During the expansion period beginning in late 1970, average weekly initial claims fell 69.5 percent in the Grand Rapids SMSA, compared to a 52.6 percent decline statewide. But, during the recession, initial claims jumped 452.6 percent in the local area. Only the Ann Arbor-Ypsilanti, Flint, and Saginaw SMSAs registered larger relative increases than Grand Rapids during that period. Interestingly, each of those areas is more heavily dependent on automobile production than Grand Rapids, and each experienced a more severe drop in manufacturing employment.

Chart GR3 shows, however, that as of the fourth quarter of 1978, initial claims in the Grand Rapids SMSA were considerably below the recession high. In fact, average weekly initial claims were down 67.4 percent in the local area, compared to 53.7 percent statewide. Nevertheless, they were still above their previous cyclical lows in the fourth quarter of 1968 and the second quarter of 1974.

Finally, Chart GR4 shows that average weekly hours of production workers in manufacturing industries in the Grand Rapids SMSA fluctuated within a rather narrow range during the 1970-78 period. During the last recession the average workweek fell only 5.3 percent. Except for the Bay City SMSA, where the average workweek did not experience a cyclical decline, that was the mildest contraction among Michigan metropolitan areas. Since the low of the first quarter of 1975, the average workweek has expanded in the area. Expansion has been relatively mild, however, hitting a recent high of 42.1 hours in the fourth quarter of 1977 and remaining close to that level since then. Throughout the recent expansion, the Grand Rapids SMSA has had the shortest workweek among the state's metropolitan areas.

Construction

Both construction employment and new building permits for private housing in the Grand Rapids SMSA exhibit cyclical behavior that conforms closely to national business cycles. Charts GR5 and GR6 show that both series declined during recessions and increased during expansion periods. The timing was different, however. New building permits led at national business cycle peaks and troughs, whereas construction employment tended to lag, especially at troughs.

The magnitude of the decline in new building permits was about the same during the 1969-70 and 1973-75 recessions, 43 percent and 46.3 percent, respectively. In all other Michigan metropolitan areas except Jackson, the contraction during the 1973-75 recession was considerably more severe than that of the late 1960s. This was also the case for the state as a whole and the nation. Interestingly, while the 1973-75 slump in this area's building permits was relatively mild—only the Jackson SMSA, with a drop in new building permits of 33.9 percent, experienced a milder contraction—the earlier decline in the Grand Rapids area had been one of the more severe among the 11 metropolitan areas in the state. In fact, the 43 percent decline in the Index of New Building Permits in this area during 1969 ranked fifth among Michigan metropolitan areas.

The expansions in new building permits have been relatively strong in the Grand Rapids SMSA. Chart GR5 shows that the peak in the index in early 1973 was above the previous cyclical high in late 1968. Preliminary data for 1978 show that the most recent level of this area's new building permits index was only 1.8 percent below its 1973 peak. Among Michigan metropolitan areas, only Jackson's Index of New Building Permits has risen above its pre-recession peak. The performance of the Grand Rapids index ranks second in this respect.

Employment in the area's construction industry fell 20.7 percent from a peak in the fourth quarter of 1973 to a low in the fourth quarter of 1976. Although that local contraction lasted three years, this was not unique to the Grand Rapids SMSA, since all metropolitan areas in Michigan experienced slumps in construction employment lasting longer than the national recession. Statewide, the cyclical decline in construction employment lasted 13 quarters. The amplitude of the contraction in the Grand Rapids SMSA was relatively small, with the peak-to-trough decline of 20.7 percent ranking behind the 13.1 percent drop in the Muskegon-Norton Shores-Muskegon Heights SMSA and the 18.2 percent falloff in the Kalamazoo-Portage SMSA.

Chart GR6 shows that since the low in late 1976, construction employment has expanded to a level in the last quarter of 1978 that is 2.7 percent above its previous cyclical peak. Thus, the Grand

Grand Rapids SMSA 115

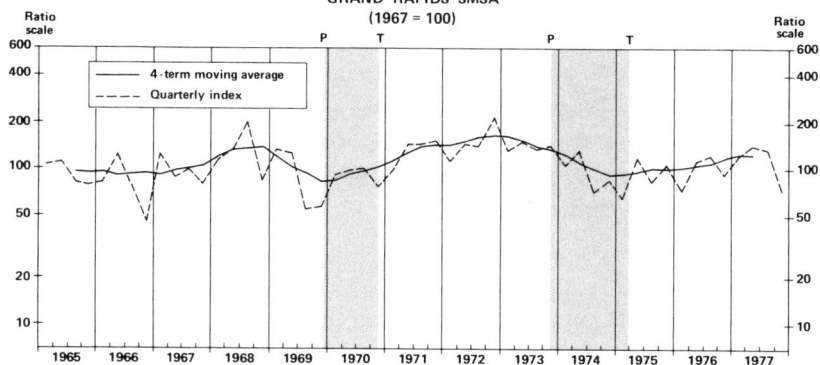

CHART G 5
INDEX OF NEW BUILDING PERMITS, PRIVATE HOUSING
GRAND RAPIDS SMSA
(1967 = 100)

Source: The W.E. Upjohn Institute. Index based on the U.S. Department of Commerce, Bureau of the Census, *Construction Reports — Housing Authorized by Building Permits and Public Contracts, C-40.*
Shaded areas indicate national recession periods as defined by the National Bureau of Economic Research, Inc. P = peak and T = trough.

CHART G 6
CONSTRUCTION EMPLOYMENT, GRAND RAPIDS SMSA
(Seasonally adjusted)

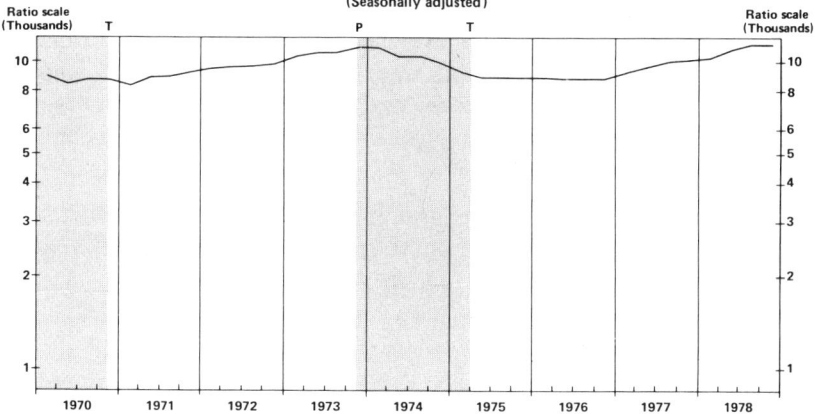

Source: Michigan Employment Security Commission.
Notes: Seasonal adjustment by the W.E. Upjohn Institute.
Shaded areas indicate national recession periods as defined by the National Bureau of Economic Research, Inc. P = peak and T = trough.

Rapids SMSA joins five other Michigan metropolitan areas—Bay City, Detroit, Flint, Kalamazoo-Portage, and Muskegon-Norton Shores-Muskegon Heights—as regions in the state where employment in the local construction industry has risen above pre-recession levels during the recent expansion. Because of the relatively strong recoveries in the Grand Rapids SMSA following the last two recessions, construction employment for the 1970-78 period grew at an annual rate of 1.5 percent, exceeding the modest 0.4 percent growth statewide.

Banking Activity

The Grand Rapids SMSA outperformed the state over the 1970-77 period in terms of growth of several banking indicators. Table GR3 shows that in current dollars the largest difference was in total deposits, which grew at an annual rate of 9.4 percent, compared to 7.3 percent statewide. In addition, the 11.3 percent growth rate in the current-dollar volume of commercial and industrial loans in the Grand Rapids SMSA was the fourth highest among the 11 metropolitan areas in Michigan. Ranking ahead of the Grand Rapids SMSA were the Saginaw SMSA with a growth rate of 14.3 percent per year, and the Ann Arbor-Ypsilanti and Muskegon SMSAs with growth rates of 13.3 percent per year. The growth rate of consumer installment loans in the Grand Rapids SMSA at 9.3 percent per year was relatively low, with only the Jackson (8.1 percent) and the Detroit (8.7 percent) SMSAs exhibiting slower growth.

The cyclical swings in loans and deposits at commercial banks in the area are revealed in Charts GR7 and GR8. In Chart GR7, the Index of Deflated Demand Deposits exhibits swings of larger magnitude than the Index of Deflated Total Deposits. The former fell 24.4 percent during the 1973-75 recession, compared to 9.1 percent for the latter. Also, the Index of Deflated Total Deposits has risen above its pre-recession peak, but the Index of Deflated Demand Deposits has not. As of the second quarter of 1978 (the latest data available), the total deposits index was 3.2 percent above its previous peak of the fourth quarter of 1973. The demand deposits index, on the other hand, was still 14.3 percent below its pre-recession peak. That situation is not unique to this area. As noted in the other sections, sluggish behavior of demand deposits in local areas and in the state as a whole is due in part to changes in banking procedures.

Grand Rapids SMSA 117

CHART G 7
INDEX OF DEFLATED TOTAL DEPOSITS AND INDEX OF DEFLATED DEMAND DEPOSITS
GRAND RAPIDS SMSA
(1972 = 100)
(Seasonally adjusted)

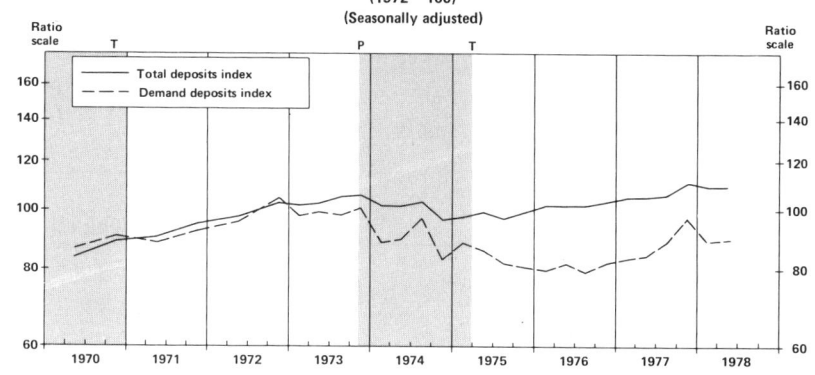

Source: The W.E. Upjohn Institute. Indices are based on data from the Federal Reserve Bank of Chicago.
Note: Shaded areas indicate national recession periods as defined by the National Bureau of Economic Research, Inc. P = peak and T = trough.

CHART G 8
COMMERCIAL BANK LOANS, GRAND RAPIDS SMSA
(Current dollars)

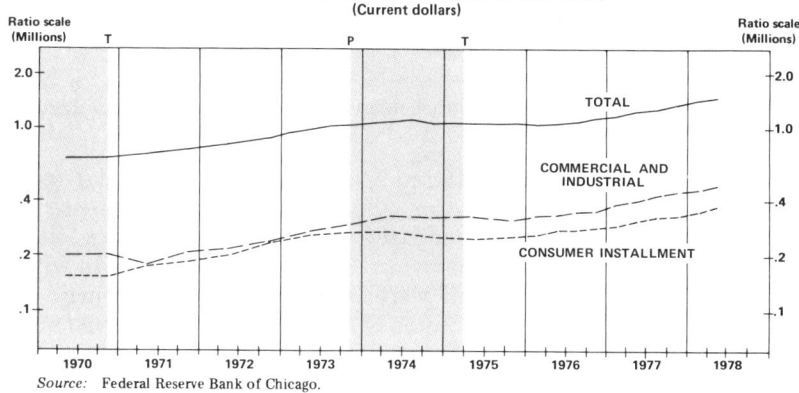

Source: Federal Reserve Bank of Chicago.
Notes: Seasonal adjustment of total loans by the W.E. Upjohn Institute. Other loans are not seasonally adjusted.
Shaded areas indicate national recession periods as defined by the National Bureau of Economic Research, Inc.
P = peak and T = trough.

TABLE GR3

Average Annual Growth Rates of Selected Commercial Banking Indicators Grand Rapids SMSA and Michigan[a]

(percent)

Indicator	Grand Rapids SMSA	Michigan
Demand deposits (current dollars)	5.0	3.7
Deflated demand deposits[b]	-2.1	-3.0
Total deposits (current dollars)	9.4	7.3
Deflated total deposits[b]	2.3	0.4
Total loans (current dollars)	8.4	7.6
Commercial and industrial loans (current dollars)	11.3	7.3
Consumer installment loans (current dollars)	9.4	9.1

[a] Except where indicated otherwise, estimated growth rates are based on log-linear trends for the 1970-78 period.

[b] Current-dollar values adjusted for changes in the U.S. Consumer Price Index.

Total loans in the area suffered a setback in current-dollar terms during the last recession, as did consumer installment and commercial and industrial loans (see Chart GR8). The latter fell 5.6 percent and the former 8.7 percent. Both contractions began with a lag during the recession and were quite brief. Consequently, the falloff in the current-dollar volume of total loans was also brief and a fairly mild 4.8 percent.

Chart GR8 shows that loan volume has expanded substantially in the area since 1975. The data for the second quarter of 1978 show total loans 40.7 percent above their recession-related low, with consumer installment loans up 49.6 percent, and commercial and industrial loans up 57.7 percent. And, of course, the current loan volume in the Grand Rapids SMSA is well above pre-recession levels, reflecting growth and expansion in current-dollar volume since the recession.

In sum, then, the Grand Rapids SMSA is characterized by cyclical behavior across key local economic processes that conforms closely to swings in national business activity, with notable differences in timing. The behavior of several economic indicators, reflecting performance in the local labor market, the area's construction sector, and local banking, suggests that this metropolitan area was the most cyclically stable area in Michigan over the period examined. Such stability appears to be linked to industrial diversification and less reliance on the automobile industry than in most other metropolitan areas in the state. However, that structure gives the area the lowest manufacturing wage rates and also one of the lowest per capita income levels among Michigan metropolitan areas.

Jackson SMSA

The Jackson SMSA is a relatively small metropolitan area encompassing the single county of Jackson, which is located in the south central part of the lower peninsula. In 1977, it had a population of 149,900. Among the 11 metropolitan areas in the state, only the Bay City SMSA is smaller. The area did experience a relatively large increase in population between 1970 and 1977, with its 4.7 percent increase ranking fourth among the state's 11 metropolitan areas, behind the increases in the Lansing-East Lansing, Ann Arbor-Ypsilanti, and Grand Rapids SMSAs.

In 1976, the Jackson SMSA generated $922 million in total personal income. As in the case of population, only the Bay City SMSA had a lower level of personal income. However, the Jackson SMSA ranked ahead of the Bay City and Grand Rapids SMSAs in terms of per capita income. In 1976, its per capita income was $6,268 compared to $6,254 in Grand Rapids and $6,116 in Bay City. Personal income in the area grew at an annual rate of 7.6 percent during the 1969-76 period, the slowest growth rate among the state's 11 metropolitan areas.

The employment distribution shown in Table J1 reveals that the Jackson SMSA is similar to the state as a whole. In 1977, 31 percent of total wage and salary employment was accounted for by manufacturing industries, with the bulk of that engaged in the production of durable goods. The major industries in the area's manufacturing sector include transportation equipment, nonelectrical machinery, and fabricated metals, with a sizable portion of the activity in those industries linked closely to the volatile automobile industry. But, during the 1970s, the local manufacturing base shrank considerably, as the greatest employment growth occurred in nonmanufacturing industries oriented to retail trade and services.

TABLE J1

Percentage Distribution of Total Wage and Salary Employment
Jackson SMSA and Michigan, 1972 and 1977

	Jackson SMSA		Michigan	
Item	1972	1977	1972	1977
Total Wage and Salary	100.0%	100.0%	100.0%	100.0%
Manufacturing	35.6	31.0	35.1	32.4
Durables	29.0	24.1	28.1	25.9
Nondurables	6.6	6.9	7.0	6.5
Nonmanufacturing	48.5	51.8	48.0	49.6
Government	15.9	17.2	16.9	18.0

Source: Michigan Employment Security Commission.

Business Conditions in the Jackson SMSA

Labor Market Conditions

The Jackson SMSA ranks with its western neighbor, the Battle Creek SMSA, as one of the two slowest growth areas in the state in terms of employment. Except for nonmanufacturing employment, each of the employment categories shown in Chart J1 grew at annual rates below those of the state as a whole (see Table J2). The performance of the local manufacturing sector was particularly weak, declining at an annual rate of 1.5 percent during the 1970-78 period. This represents the steepest downtrend among the 11 metropolitan areas in Michigan, joining the Battle Creek, Detroit, and Muskegon-Norton Shores-Muskegon Heights SMSAs as areas experiencing negative growth in manufacturing employment over the nine-year period.

Chart J1 shows that manufacturing employment in the area experienced a long contraction that began in the third quarter of 1973 and did not end until the second quarter of 1976—the longest slump among the state's metropolitan areas. Only the Battle Creek SMSA, with a decline lasting nine quarters, came close to the Jackson SMSA in terms of duration.

In addition to its long duration, the slump in the area's manufacturing employment was also quite severe, measuring 25.5 percent from peak to trough. That equals the decline in the Flint

Jackson SMSA

CHART J 1
WAGE AND SALARY EMPLOYMENT, JACKSON SMSA
(Seasonally adjusted)

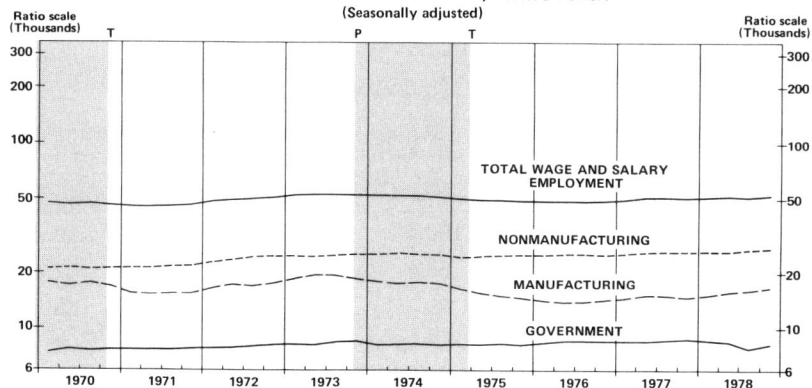

Source: Michigan Employment Security Commission.
Notes: Seasonal adjustment by the W.E. Upjohn Institute.
Shaded areas indicate national recession periods as defined by the National Bureau of Economic Research, Inc.
P = peak and T = trough.

CHART J 2
UNEMPLOYMENT RATE
JACKSON SMSA
(Seasonally adjusted)

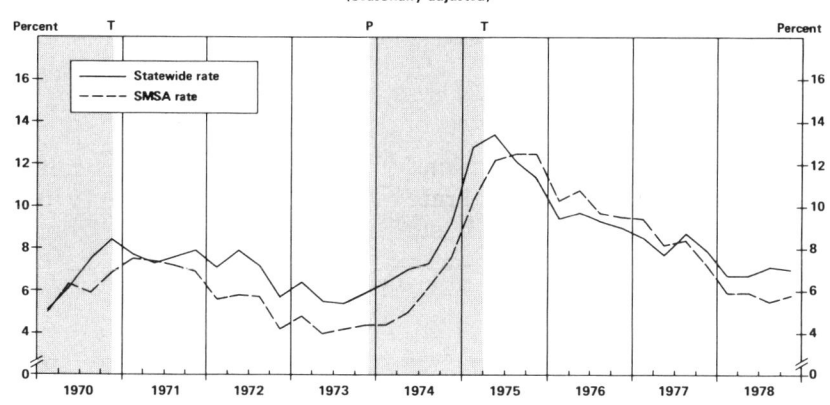

Source: Michigan Employment Security Commission.
Note: Seasonal adjustment by the W.E. Upjohn Institute.
Shaded areas indicate national recession periods as defined by the National Bureau of Economic Research, Inc.
P = peak and T = trough.

TABLE J2

Average Annual Growth Rates of
Selected Labor Market Indicators
Jackson SMSA and Michigan[a]

(percent)

Indicator	Jackson SMSA	Michigan
Total wage and salary employment	1.1	1.8
Manufacturing employment	-1.5	0.2
Nonmanufacturing employment	2.8	2.6
Government employment	1.6	3.0
Civilian labor force	1.6	2.0
Unemployment rate	4.3	2.8
Average weekly initial claims for UI[b]...	7.5	7.3
Average workweek, production workers, mfg.[b]	0.7	0.2

[a] Except where indicated otherwise, estimated growth rates are based on log-linear trends for the 1970-78 period.

[b] Computed for the 1968-78 period.

SMSA and nearly equals the drop of 26 percent in the Ann Arbor-Ypsilanti SMSA, the most severe decline among the metropolitan areas in the state. However, unlike those two areas, where manufacturing employment recovered and then expanded to levels above pre-recession peaks, the recovery in manufacturing employment in the Jackson SMSA was quite weak. At the end of 1978, it was only 17.8 percent above the low in 1976 and still 14 percent below its pre-recession peak.

As a result of a weak manufacturing sector, total wage and salary employment in the area has also performed poorly compared to other metropolitan areas in the state, with the possible exception of the Battle Creek SMSA. During the 1973-75 recession, total wage and salary employment fell 8.3 percent, second only to the 13.4 percent falloff in the Flint SMSA. But, again the recovery was weak, as total wage and salary employment rose just 8.8 percent from a low in the third quarter of 1976 to the fourth quarter of 1978. Moreover, that increase still left the area's total wage and salary employment just below its pre-recession peak. Thus, the Jackson SMSA is the only metropolitan area in Michigan in which wage and salary employment did not exceed its pre-recession peak by the end of 1978.

The unemployment rate and average weekly initial claims for unemployment insurance also reflect the relatively weak performance of this area's economy. Over the 1970-78 period, the jobless rate in the Jackson SMSA drifted upward at an annual rate of 4.3 percent, compared to 2.8 percent for the state as a whole. In addition, over a slightly longer period of time—1968-1978—average weekly initial claims in the area grew at an annual rate of 7.5 percent. As Table J2 shows, the comparable growth rate statewide is 7.3 percent.

Charts J2 and J3 show that the seasonally-adjusted, local unemployment rate and average weekly initial claims in the area experienced sizable increases during the 1973-75 recession. The jobless rate rose from 4.0 percent in the second quarter of 1973 to 12.5 percent in the fourth quarter of 1975. That runup in the unemployment rate was exceeded only by the increases that occurred in the Ann Arbor-Ypsilanti and Flint SMSAs. As Chart J2 shows, the local unemployment rate remained above the statewide average from late 1975 through mid-1977. Since then, the local jobless rate has moved well below the statewide average, with the former averaging 5.6 percent and the latter 7.0 percent during the fourth quarter of 1978.

However, while initial claims for unemployment insurance rose 282 percent during the last recession, that was below the increase of 301.1 percent recorded statewide. Besides, that local increase was the fourth mildest among the state's metropolitan areas, ranking behind the rise in the Muskegon, Lansing-East Lansing, and Kalamazoo-Portage SMSAs. As Chart J3 shows, the decline in average weekly initial claims in the Jackson SMSA during the recent expansion period has brought the level down considerably, as they fell 59.9 percent from their high in the third quarter of 1975. At the end of 1978, average weekly initial claims were still 53 percent above their pre-recession low.

The average workweek of production workers in the manufacturing sector is shown in Chart J4. Note that average weekly hours began declining before the two national recessions started. Interestingly, in the Jackson SMSA, the relative decline during the 1969-70 recession was larger than that of the 1973-75 period, 8.9 percent compared to 7.6 percent. This is similar to the behavior in the Saginaw SMSA, where the shrinkage in the workweek was

Jackson SMSA

CHART J 3
AVERAGE WEEKLY INITIAL CLAIMS FOR UNEMPLOYMENT INSURANCE, STATE PROGRAMS
JACKSON SMSA
(Seasonally adjusted)

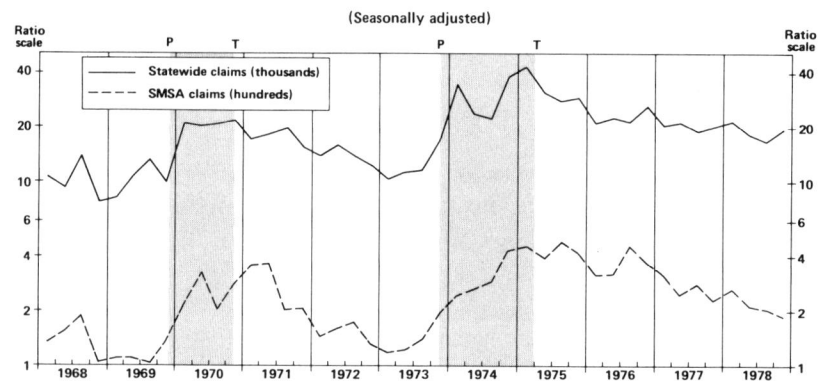

Source: The W.E. Upjohn Institute, based on data from the Michigan Employment Security Commission and the U.S. Department of Labor.
Shaded areas indicate national recession periods as defined by the National Bureau of Economic Research, Inc.
P = peak and T = trough.

CHART J 4
AVERAGE WEEKLY HOURS OF PRODUCTION WORKERS IN MANUFACTURING INDUSTRIES
JACKSON SMSA
(Seasonally adjusted)

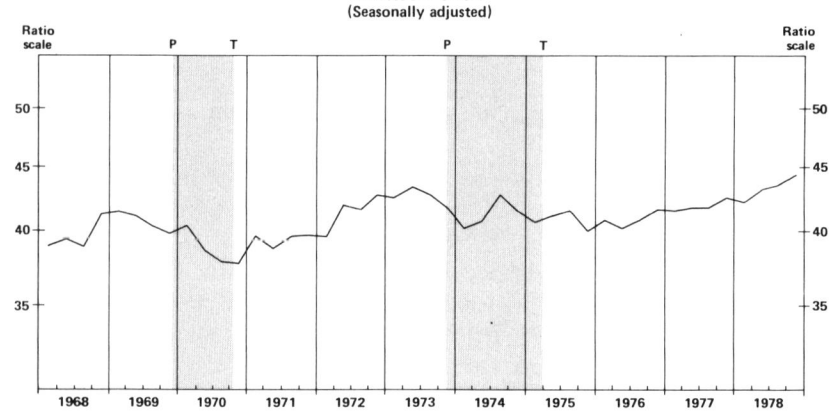

Source: Michigan Employment Security Commission.
Note: Seasonal adjustment by the W.E. Upjohn Institute.
Shaded areas indicate national recession periods as defined by the National Bureau of Economic Research, Inc.
P = peak and T = trough.

Jackson SMSA

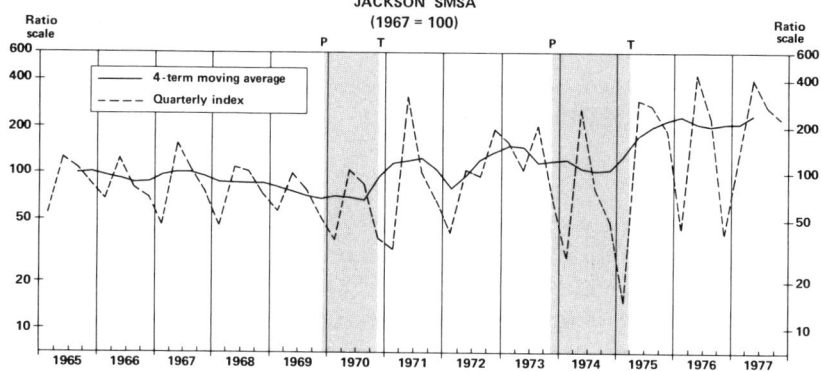

CHART J 5
INDEX OF NEW BUILDING PERMITS, PRIVATE HOUSING
JACKSON SMSA
(1967 = 100)

Source: The W.E. Upjohn Institute. Index based on the U.S. Department of Commerce, Bureau of the Census, *Construction Reports — Housing Authorized by Building Permits and Public Contracts, C-40.*

Shaded areas indicate national recession periods as defined by the National Bureau of Economic Research, Inc.
P = peak and T = trough.

CHART J 6
CONSTRUCTION EMPLOYMENT, JACKSON SMSA
(Seasonally adjusted)

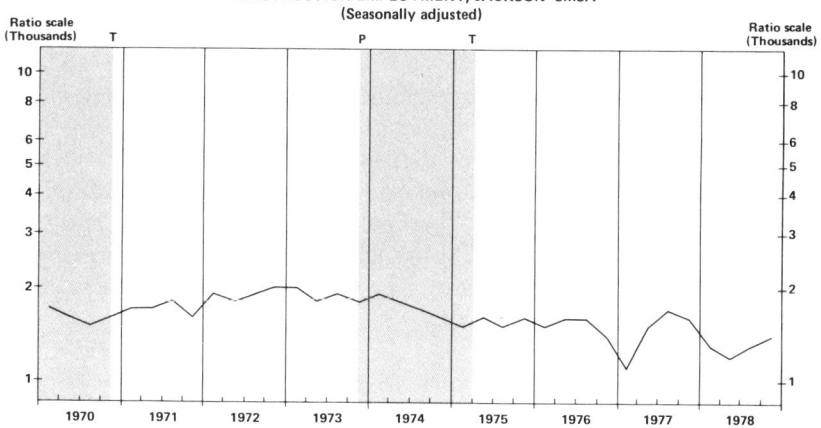

Source: Michigan Employment Security Commission.
Notes: Seasonal adjustment by the W.E. Upjohn Institute.
Shaded areas indicate national recession periods as defined by the National Bureau of Economic Research, Inc.
P = peak and T = trough.

larger during the 1969-70 recession. In the Jackson SMSA, the average workweek has exhibited fairly strong recoveries after recessions, rising in each expansion to levels above pre-recession peaks. In the last quarter of 1978, the workweek in the area averaged 44.3 hours, up from its previous cyclical high of 43.3 hours in the second quarter of 1973.

Construction

Two indicators of construction activity in the Jackson SMSA, shown in Charts J5 and J6, provide conflicting signals about the vigor of such activity in the area. New building permits for private housing, shown in index form (1967=100) in Chart J5, performed better in terms of growth and cyclical swings in this area than in any other metropolitan area in the state. From 1965 to 1977, they grew at an annual rate of 8.4 percent. Over that same period, new building permits declined 2 percent per year for Michigan as a whole and rose at a modest annual rate of 1.2 percent nationwide. The growth in the Jackson SMSA was also far above that of other metropolitan areas in the state. In fact, only three other areas experienced positive growth over those 13 years: Bay City (3.2 percent), Grand Rapids (1.6 percent), and Muskegon-Norton Shores-Muskegon Heights (0.4 percent).

The cyclical downswings in new building permits in the area were relatively mild, falling 33.3 percent and 33.9 percent during the 1969-70 and 1973-75 recessions, respectively. The latter was the mildest among the state's 11 metropolitan areas, and considerably milder than the 56.2 percent drop in the state as a whole and the 69.1 percent decline nationwide. As Chart J5 shows, the expansion in new building permits locally has been quite strong since late 1974. In fact, the Jackson SMSA is the only metropolitan area in Michigan in which the moving average Index of New Building Permits has risen above its pre-recession peak during the recent expansion.

However, employment in the local construction industry did not exhibit similar strength. Chart J6 shows that construction employment experienced a long cyclical contraction that began in early 1973 and did not end until the first quarter of 1977. Except for the Battle Creek SMSA, that was the longest downswing among the metropolitan areas in Michigan. It was also the most severe, with employment dropping 45.0 percent from peak to trough. The recovery of construction employment has been weak in the Jackson SMSA. Chart J6 shows that the area's construction employment is still well below its pre-recession peak. In fact, quarterly employment levels in 1978 were below those of the second, third, and fourth quarters of 1977. Over the entire 1970-78 period, construction employment in the area has fallen at an annual rate of 2.8 percent. This negative growth for construction employment is certainly in sharp contrast to the fairly rapid growth in new building permits for private housing in the area.

Banking Activity

The performance of loans and deposits at commercial banks in the Jackson SMSA was not markedly different from the state as a whole over the 1970-77 period. Table J3 shows that total deposits and demand deposits in the local area grew at slightly faster rates than the state aggregates, but loans grew at a somewhat slower pace locally. However, the 7.9 percent rate of growth in the current-dollar volume of total deposits in the Jackson SMSA was a relatively slow rate compared to most other metropolitan areas in Michigan. Only the Bay City, Detroit, and Lansing-East Lansing SMSAs exhibited slower deposit growth. Those same three areas also experienced slower growth in total loans.

TABLE J3

Average Annual Growth Rates of
Selected Commercial Banking Indicators
Jackson SMSA and Michigan[a]

(percent)

Indicator	Jackson SMSA	Michigan
Demand deposits (current dollars)	3.8	3.7
Deflated demand deposits[b]	-2.8	-3.0
Total deposits (current dollars)	7.9	7.3
Deflated total deposits[b]	0.9	0.4
Total loans (current dollars)	7.1	7.6
Commercial and industrial loans (current dollars)	6.1	7.3
Consumer installment loans (current dollars)	8.1	9.1

[a]Except where indicated otherwise, estimated growth rates are based on log-linear trends for the 1970-78 period.

[b]Computed for the 1968-78 period.

130   Jackson SMSA

The two loan categories shown in Table J3 also reflect slow growth in the Jackson SMSA. The 6.1 percent annual growth rate for commercial and industrial loans only surpassed the very modest growth of 3.3 percent in the Bay City SMSA and the 5.7 percent rate in the Detroit SMSA. In addition, the 8.1 percent annual increase in consumer installment loans in the Jackson SMSA ranked tenth among the 10 metropolitan areas in the state for which comparable data were available.[1]

The cyclical swings in demand deposits and total deposits at commercial banks in this area are revealed clearly in Chart J7. Both indexes, which are based on current-dollar values adjusted for price changes, started to decline before the 1973-75 national recession began. This leading behavior was evident in all Michigan metropolitan areas. In the Jackson SMSA the contraction in both deposit indexes continued for several quarters after the slump in national business activity ended. Therefore, like other metropolitan areas in Michigan, the cyclical downswings in the two deflated deposits series for this area were relatively long.

From peak to trough the demand deposits index declined 26.4 percent and the total deposits index fell 10.3 percent. Chart J7 shows that, while both indexes have recovered from their recession lows, neither has moved above its pre-recession peak. In the second quarter of 1978, the demand deposits index was still 18.9 percent below its 1973 high and the total deposits index was 6 percent lower. Thus, the Jackson SMSA joins six other metropolitan areas in Michigan in which the index of price adjusted total deposits in commercial banks was below the pre-recession peak as of mid-1978.[2]

Chart J8 shows that the current-dollar volume of loans at commercial banks in this area exhibited a degree of cyclical sensitivity during the 1970-78 period. On a current-dollar seasonally adjusted basis, total loans fell 6.4 percent from a high in the first quarter of 1974 to a low two years later. That is the third most severe decline in loan volume among the 11 metropolitan areas in the state. Only the Battle Creek and Detroit SMSAs experienced cyclical declines of greater magnitude, 9.8 percent and 7 percent, respectively. Although the behavior of consumer installment loans and commercial and industrial loans in the area is not as clear,

---

1. Battle Creek is not strictly comparable because its banking data began in 1972 compared to 1970 for all other metropolitan areas in Michigan.
2. The other SMSAs are Battle Creek, Bay City, Detroit, Kalamazoo-Portage, Lansing-East Lansing, and Muskegon.

Jackson SMSA

CHART J 7
INDEX OF DEFLATED TOTAL DEPOSITS AND INDEX OF DEFLATED DEMAND DEPOSITS
JACKSON SMSA
(1972 = 100)
(Seasonally adjusted)

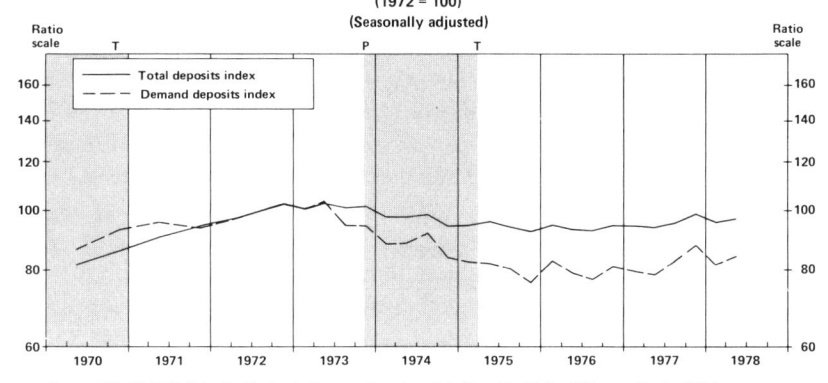

Source: The W.E. Upjohn Institute. Indices are based on data from the Federal Reserve Bank of Chicago.
Note: Shaded areas indicate national recession periods as defined by the National Bureau of Economic Research, Inc. P = peak and T = trough.

CHART J 8
COMMERCIAL BANK LOANS, JACKSON SMSA
(Current dollars)

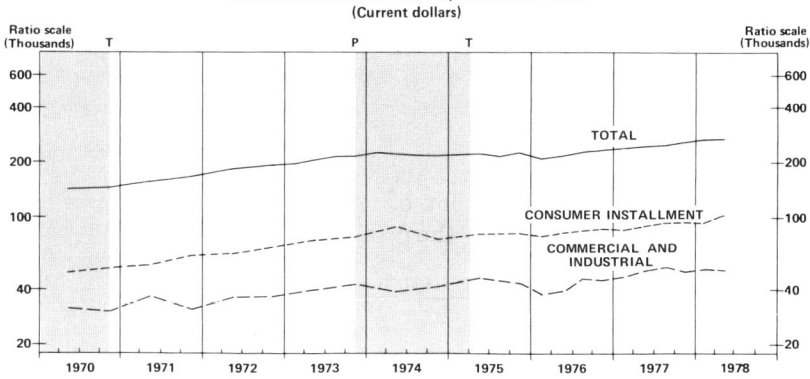

Source: Federal Reserve Bank of Chicago.
Notes: Seasonal adjustment of total loans by the W.E. Upjohn Institute. Other loans are not seasonally adjusted.
Shaded areas indicate national recession periods as defined by the National Bureau of Economic Research, Inc. P = peak and T = trough.

several quarters of weakness are apparent in each series during 1974 and 1975. Of course, like other metropolitan areas in Michigan, the current-dollar volume of loans in the Jackson SMSA was well above pre-recession levels as of mid-1978.

In summary, overall economic activity in the Jackson SMSA has been relatively weak over roughly the last ten years. The area has experienced slow growth in wage and salary employment, attributable largely to the secular decline in its manufacturing base. Certainly, the 1973-75 national recession had a severe impact on the area. And, based on the measures covered in this study, the economic recovery and expansion since the recession have not been vigorous. In fact, compared to other metropolitan areas in the state, the business expansion in the last few years has been rather weak. Consequently, the Jackson SMSA and the neighboring Battle Creek SMSA displayed two of the weakest economic performances among metropolitan areas in Michigan over the period examined.

Kalamazoo-Portage SMSA

This metropolitan area is made up of Kalamazoo and Van Buren counties, with the Battle Creek SMSA bordering on the east and Lake Michigan to the west. In 1977, the Kalamazoo-Portage SMSA had a population of 268,100, making it the fifth largest metropolitan area in the state, ranking just ahead of the Ann Arbor-Ypsilanti SMSA. About 77 percent of the metropolitan population resides in Kalamazoo County, which contains the urban center cities of Kalamazoo and Portage. Between 1970 and 1977 the two-county area's population increased by 4 percent, exceeding the statewide rise of 2.8 percent. However, the population of Kalamazoo County rose only 2.3 percent, while the population of the more rural Van Buren County increased 10.1 percent.

Total personal income in the Kalamazoo-Portage SMSA amounted to $1,719 million in 1976. That represented the sixth largest income among the state's 11 metropolitan areas, ranking just below the Ann Arbor-Ypsilanti SMSA, which had a personal income level of $1,790 million in that year. With a per capita income of $6,510 in 1976, the Kalamazoo-Portage SMSA ranked fifth, somewhat below the statewide level of $6,757, but still above the national level of $6,396. From 1969 to 1976, personal income in the Kalamazoo-Portage SMSA grew at an average annual rate of 9 percent, the fifth highest growth rate among the 11 metropolitan areas in the state. It should be noted that this metropolitan area is not a high-wage area. In fact, the average wage in manufacturing industries in the Kalamazoo-Portage SMSA is below the statewide average, but it ranks ahead of other lower-wage areas like Bay City, Grand Rapids, Jackson, and Muskegon. This, of course, is due in part to the fact that the automobile industry accounts for a relatively small share of the manufacturing base in the area.

Table K1 shows that in terms of manufacturing, nonmanufacturing, and government, the distribution of employment in the Kalamazoo-Portage SMSA is quite similar to the state as a whole. There is one outstanding difference between the two, however. Manufacturing employment in the state is heavily concentrated in durable goods industries, which accounted for 25.9 percent of total wage and salary and 80 percent of manufacturing employment in 1977. In contrast, durable goods industries accounted for slightly less than one-half of all manufacturing jobs in the Kalamazoo-Portage SMSA. And among the state's 11 metropolitan areas, the Kalamazoo-Portage SMSA is the least dependent on cyclically sensitive hard-goods industries. The major manufacturing industry in the area is the chemicals and allied products group, primarily pharmaceutical production. The Upjohn Company is the largest single employer in the area and has been characterized by a high degree of cyclical stability, historically.

TABLE K1

Percentage Distribution of Total Wage and Salary Employment
Kalamazoo-Portage SMSA and Michigan, 1972 and 1977

Item	Kalamazoo-Portage SMSA 1972	Kalamazoo-Portage SMSA 1977	Michigan 1972	Michigan 1977
Total Wage and Salary	100.0%	100.0%	100.0%	100.0%
Manufacturing	37.6	33.8	35.1	32.4
Durables	18.1	16.4	28.1	25.9
Nondurables	19.5	17.5	7.0	6.5
Nonmanufacturing	42.8	46.3	48.0	49.6
Government	19.6	19.9	16.9	18.0

Source: Michigan Employment Security Commission.

Like the other metropolitan areas in the western part of lower Michigan, business activity in the Kalamazoo-Portage SMSA is not heavily concentrated in automobile production. In 1977, the transportation equipment industry accounted for only 11.4 percent of manufacturing employment in the area, well below the 34.7 percent share statewide. Fabricated metal production accounted for 14 percent of the area's manufacturing employment in 1977, with the bulk of that activity linked to the automobile industry. Nevertheless, the Kalamazoo-Portage SMSA has one of the most diversified manufacturing bases among the metropolitan areas in

Michigan with 16.5 percent of the local industrial workforce employed in the production of paper and paper products. Almost one quarter of the statewide wage and salary workers in paper production had jobs located within this SMSA in 1977.

In addition, a large state university, Western Michigan University, and several smaller colleges, Kalamazoo College, Nazareth College, and Kalamazoo Valley Community College, are located in this metropolitan area. The heavy concentration of higher educational institutions makes the Kalamazoo-Portage SMSA similar to the Lansing-East Lansing and Ann Arbor-Ypsilanti SMSAs in this respect.

Business Conditions in the Kalamazoo-Portage SMSA

Labor Market Conditions

Key indicators of employment and unemployment activity reveal a relatively high degree of cyclical stability and fairly strong growth in this area. In fact, over the 1970-78 period, the Kalamazoo-Portage SMSA experienced the mildest cyclical swings among the state's 11 metropolitan areas. Local manufacturing and total wage and salary employment did suffer setbacks during the 1973-75 recession, but they were not severe. Manufacturing employment fell 13.6 percent from a high in the third quarter of 1973 to a low in the second quarter of 1975. The absolute decline in total wage and salary employment was very brief, only two quarters, and amounted to only a 3.2 percent falloff, the mildest contraction among Michigan metropolitan areas.

Chart K1 shows that the downswing in manufacturing employment during the recession was offset to a large extent by increases in nonmanufacturing and government employment in the area. While the size and stability of the nonmanufacturing and government sectors contribute to the overall stability of wage and salary employment in this area, it should be noted that the manufacturing sector in the Kalamazoo-Portage SMSA has exhibited less severe cyclical swings compared to other metropolitan areas in Michigan. As noted above, this is due to the fact that the area does not specialize in durable goods production in general, nor in automobile production in particular, to the extent evident in most other Michigan metropolitan areas.

Chart K1 shows that manufacturing employment increased throughout the recent expansion period. By the end of 1978, it was

136 Kalamazoo-Portage SMSA

CHART K 1
WAGE AND SALARY EMPLOYMENT, KALAMAZOO-PORTAGE SMSA
(Seasonally adjusted)

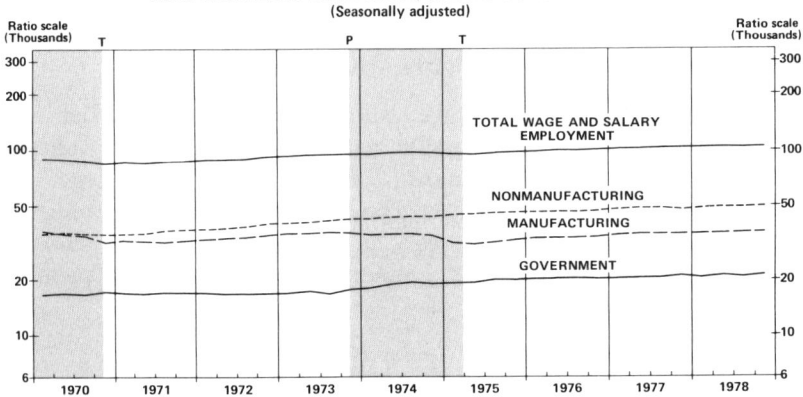

Source: Michigan Employment Security Commission.
Notes: Seasonal adjustment by the W.E. Upjohn Institute.
Shaded areas indicate national recession periods as defined by the National Bureau of Economic Research, Inc.
P = peak and T = trough.

CHART K 2
UNEMPLOYMENT RATE
KALAMAZOO - PORTAGE SMSA
(Seasonally adjusted)

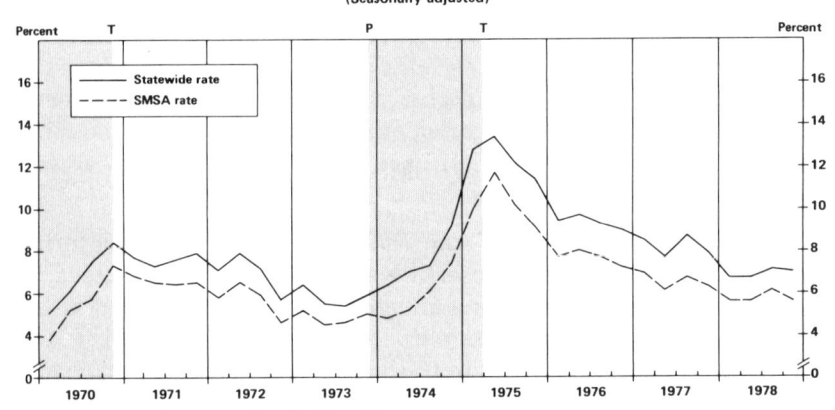

Source: Michigan Employment Security Commission.
Note: Seasonal adjustment by the W.E. Upjohn Institute.
Shaded areas indicate national recession periods as defined by the National Bureau of Economic Research, Inc.
P = peak and T = trough.

18.1 percent above its recession low and 1.9 percent above its prerecession peak. Table K2 shows that, except for government employment, the major employment categories grew more rapidly in the Kalamazoo-Portage SMSA than statewide. The growth of nonmanufacturing employment was especially strong, 4.6 percent per year over the nine-year period compared to 2.6 percent for the state as a whole. That local growth rate ranks second only to the 5.2 percent annual increase in the Ann Arbor-Ypsilanti SMSA. The growth of nonmanufacturing employment in the Kalamazoo-Portage SMSA has resulted from expansions in the local retail and service industries, which represent the two major growth sectors in the area.

The local unemployment rate, shown in Chart K2, remained below the state average throughout the 1970-78 period. As Chart K2 shows, the local jobless rate did exhibit cyclical swings which conformed to natural business cycle patterns. During the 1973-75 recession, the unemployment rate rose from a low of 4.5 percent to a high of 11.7 percent. Despite the sizable runup in the unemployment rate, the high in the Kalamazoo-Portage SMSA was below the statewide high of 13.4 percent. Moreover, this was the only metropolitan area in Michigan to have a cyclical high below 12 percent.

TABLE K2

Average Annual Growth Rates of
Selected Labor Market Indicators
Kalamazoo-Portage SMSA and Michigan[a]

(percent)

Indicator	Kalamazoo-Portage SMSA	Michigan
Total wage and salary employment	2.7	1.8
Manufacturing employment	0.6	0.2
Nonmanufacturing employment	4.6	2.6
Government employment	2.7	3.0
Civilian labor force	2.8	2.0
Unemployment rate	2.6	2.8
Average weekly initial claims for UI[b]	9.4	7.3
Average workweek, production workers, mfg.	-0.3	0.6

[a] Except where indicated otherwise, estimated growth rates are based on log-linear trends for the 1970-78 period.

[b] Computed for the 1968-78 period.

Since the recession, both the local and statewide unemployment rates have fallen substantially. In the last quarter of 1978, the seasonally-adjusted jobless rate in the Kalamazoo-Portage SMSA averaged 5.5 percent. Only the Ann Arbor-Ypsilanti, Grand Rapids, and Saginaw SMSAs recorded lower jobless rates for any quarter during 1978. Chart K2 shows that the differential between the state and local rates was maintained throughout the recent expansion.

For the entire nine-year period, the unemployment rate in the Kalamazoo-Portage SMSA did drift upward at an annual rate of 2.6 percent, just below the growth of 2.8 percent statewide. It should be noted, however, that except for the Bay City and Grand Rapids SMSAs, all metropolitan areas in Michigan exhibited an upward trend in the unemployment rate over these nine years. The same behavior was evident nationwide, with the aggregate unemployment rate moving up at an annual rate of 4.3 percent, which exceeds the growth in all Michigan metropolitan areas except Jackson.

Despite the relatively good performance of employment and unemployment in the Kalamazoo-Portage SMSA, initial claims for unemployment insurance grew at a relatively rapid rate. Table K2 shows that over the 1968-78 period, average weekly initial claims grew at an annual rate of 9.4 percent as compared to 7.3 percent statewide. Moreover, this upward trend, which is quite evident in Chart K3, was the third highest among the state's 11 metropolitan areas. Only the Battle Creek and Ann Arbor-Ypsilanti SMSAs, with growth rates of 10 percent and 11.1 percent, respectively, experienced a greater upward drift in initial claims.

However, the cyclical swings in average weekly initial claims were relatively mild in the Kalamazoo-Portage SMSA. They rose 177.6 percent during the 1969-70 recession and 258 percent during the more severe 1973-75 slump. In each instance, the increases in this area ranked third lowest among the 11 metropolitan areas in Michigan and were below the statewide increase of 193.2 percent and 301.1 percent for the two recession periods, respectively. But the declines during the expansion periods were also relatively mild. As Chart K3 shows, the 1973 low in average weekly initial claims was well above the 1968 low. By the end of 1978, after more than three years of decline, average weekly initial claims in this area were still 62.3 percent above their pre-recession low. Thus, the relatively stable behavior of initial claims, reflected in the moderate runups during recessions and even more moderate falloffs during expansions, has given rise to a fairly strong upward drift over the

Kalamazoo-Portage SMSA 139

CHART K 3
AVERAGE WEEKLY INITIAL CLAIMS FOR UNEMPLOYMENT INSURANCE, STATE PROGRAMS
KALAMAZOO-PORTAGE SMSA
(Seasonally adjusted)

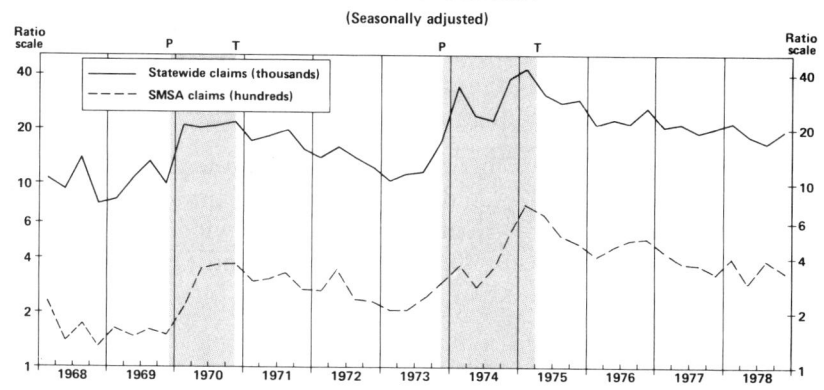

Source: The W.E. Upjohn Institute, based on data from the Michigan Employment Security Commission and the U.S. Department of Labor.
Shaded areas indicate national recession periods as defined by the National Bureau of Economic Research, Inc.
P = peak and T = trough.

CHART K 4
AVERAGE WEEKLY HOURS OF PRODUCTION WORKERS IN MANUFACTURING INDUSTRIES
KALAMAZOO – PORTAGE SMSA
(Seasonally adjusted)

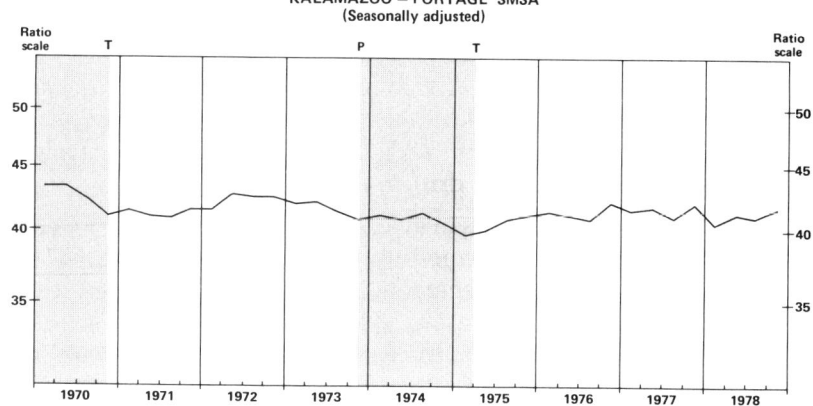

Source: Michigan Employment Security Commission.
Note: Seasonal adjustment by the W.E. Upjohn Institute.
Shaded areas indicate national recession periods as defined by the National Bureau of Economic Research, Inc.
P = peak and T = trough.

ten-year period. In many areas where the cyclical swings have been much more pronounced, the Flint and Saginaw SMSAs, for example, the growth rate over the entire period has been much lower.

The average workweek of production workers in manufacturing in the local area, shown in Chart K4, is also fairly stable. During the 1973-75 recession, the average workweek in the area fell 7 percent, ranking behind the Bay City, Battle Creek, and Grand Rapids SMSAs in terms of mildness. As would be expected, the areas in Michigan less dependent on the automobile industry exhibit much less cyclical volatility in hours worked in local manufacturing industries. During the nine-year period, the average workweek moved up and down with the business cycle, varying between a high of 43.3 hours in early 1970 and a low of 39.7 hours in the first quarter of 1975. Chart K4 shows that since that time, average weekly hours have expanded in the area but failed to reach the high recorded during early 1970. Consequently, this local series exhibits a modest downtrend of 0.3 percent per year over the period, compared to a slight upward growth of 0.6 percent statewide. Over the period examined, the Kalamazoo-Portage SMSA was the only metropolitan area in Michigan to display negative growth in the average workweek.

Construction

Building activity in the Kalamazoo-Portage SMSA, reflected in new building permits for private housing and construction employment, exhibits a high degree of cyclical sensitivity. Chart K5 shows that the moving average of new building permits in the area (the solid line) led national business cycle peaks and troughs by considerable margins. As a result, new building permits rose throughout the two national recessions, giving the impression of countercyclical behavior. But the overall behavior of new building permits is not countercyclical, as the expansion movements demonstrate. The upswings during recessions are simply the result of long lead times that characterized this local series during the period.

The magnitudes of the cyclical swings were relatively large. New building permits dropped 58.3 percent and 75.3 percent in the 1967-69 and 1972-73 periods, respectively. In the late 1960s, only the Lansing-East Lansing SMSA registered a more severe decline. In addition, the downswing in the Kalamazoo-Portage SMSA in the late 1960s was longer and considerably greater than the 21.6

Kalamazoo-Portage SMSA 141

CHART K 5
INDEX OF NEW BUILDING PERMITS, PRIVATE HOUSING
KALAMAZOO - PORTAGE SMSA
(1967 = 100)

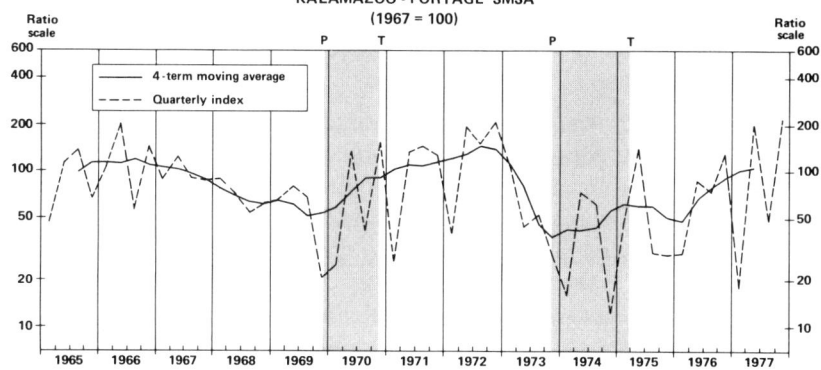

Source: The W.E. Upjohn Institute. Index based on the U.S. Department of Commerce, Bureau of the Census, *Construction Reports — Housing Authorized by Building Permits and Public Contracts, C-40.*

Shaded areas indicate national recession periods as defined by the National Bureau of Economic Research, Inc. P = peak and T = trough.

CHART K 6
CONSTRUCTION EMPLOYMENT, KALAMAZOO-PORTAGE SMSA
(Seasonally adjusted)

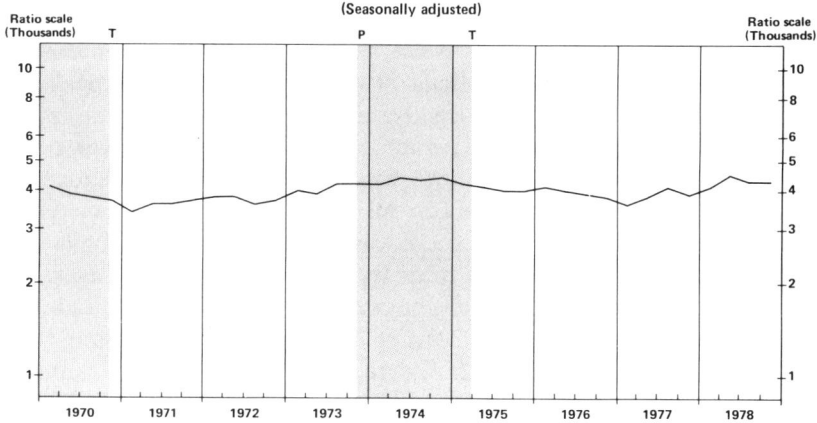

Source: Michigan Employment Security Commission.
Notes: Seasonal adjustment by the W.E. Upjohn Institute.

Shaded areas indicate national recession periods as defined by the National Bureau of Economic Research, Inc. P = peak and T = trough.

percent fall recorded for Michigan as a whole and the 24.5 percent decline in new building permits nationwide. However, while the 75.3 percent drop in the local series during the last recession was not considerably greater than the 69.1 percent decline nationwide, it was well above the 56.2 percent falloff for the state as a whole.

As Chart K5 shows, the expansion phases have been fairly strong, with new building permits rising well above their cyclical lows. Preliminary data for 1978 put the most recent high in the moving average 212 percent above the recession low. Despite this sizable increase, the Index of New Building Permits in the Kalamazoo-Portage SMSA has not risen above the 1972 peak. A similar situation exists nationwide, statewide, and in all other Michigan metropolitan areas except the Jackson SMSA.

Chart K6 shows that employment in the local construction industry lagged at national business cycle peaks and troughs. The lag at the two troughs is characteristic of this series in nearly all Michigan metropolitan areas, with the lag after the 1975 business cycle trough being particularly long. The peak in construction employment did not occur until midway through the recession, and this lag is attributable to unique local factors, mainly an increase in nonresidential building in 1974, that boosted employment. Compared to other metropolitan areas in the state, the peak-to-trough fall in construction employment of 18.2 percent was mild. Only the Muskegon SMSA experienced a less severe downswing. As Chart K6 shows, construction employment in the Kalamazoo-Portage SMSA rose substantially during 1977 and 1978. The recent high in the second quarter of 1978 was about 2.3 percent above the peak in 1974. Thus, this area joins five other metropolitan areas in the state in which construction employment in 1978 moved above pre-recession highs.[1] For the entire 1970-78 period, construction employment in the Kalamazoo-Portage SMSA grew at an annual rate of 1.2 percent, ranking fourth highest among the 11 Michigan metropolitan areas and exceeding the 0.4 percent growth statewide.

Banking Activity

The Kalamazoo-Portage SMSA outperformed the state as a whole in terms of growth and cyclical swings in key indicators of banking activity. Table K3 shows that the growth of demand and total deposits at local commercial banks was more rapid than statewide during the 1970-77 period. However, while the 6.9 percent annual

---

1. The other areas are Bay City, Detroit, Flint, Grand Rapids, and Muskegon.

growth rate for demand deposits ranks first among the 11 metropolitan areas in Michigan, the 8.9 percent growth of total deposits ranks a surprisingly low sixth.

TABLE K3

Average Annual Growth Rates of
Selected Commercial Banking Indicators
Kalamazoo-Portage SMSA and Michigan[a]

(percent)

Indicator	Kalamazoo-Portage SMSA	Michigan
Demand deposits (current dollars)	6.9	3.7
Deflated demand deposits[b]	c	-3.0
Total deposits (current dollars)	8.9	7.3
Deflated total deposits[b]	2.0	0.4
Total loans (current dollars)	9.4	7.6
Commercial and industrial loans (current dollars)	10.8	7.3
Consumer installment loans (current dollars)	12.6	9.1

[a]Except where indicated otherwise, estimated growth rates are based on log-linear trends for the 1970-78 period.

[b]Current-dollar values adjusted for changes in the U.S. Consumer Price Index.

[c]Less than one-tenth of one percent.

Over the 1970-77 period, the current-dollar volume of loans expanded at an annual rate of 9.4 percent, fourth highest among the 11 metropolitan areas in Michigan. Consumer installment loans grew at a relatively rapid rate of 12.6 percent over the eight-year period, ranking only behind the 14.5 percent growth in the Saginaw SMSA. But the 10.8 percent growth in the current-dollar volume of commercial and industrial loans ranks fifth among the 11 metropolitan areas, behind the Saginaw, Ann Arbor-Ypsilanti, Muskegon-Norton Shores-Muskegon Heights, and Grand Rapids SMSAs.

Chart K7 shows the behavior of total deposits and demand deposits over the entire period. Both series have been adjusted for

144 Kalamazoo-Portage SMSA

CHART K 7
INDEX OF DEFLATED TOTAL DEPOSITS AND INDEX OF DEFLATED DEMAND DEPOSITS
KALAMAZOO - PORTAGE SMSA
(1972 = 100)
(Seasonally adjusted)

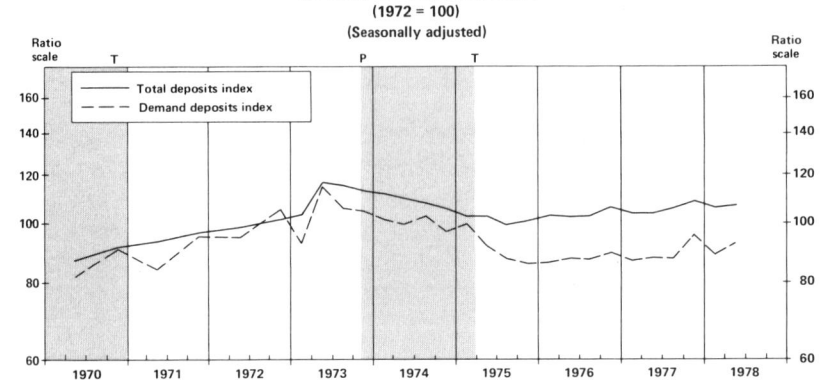

Source: The W.E. Upjohn Institute. Indices are based on data from the Federal Reserve Bank of Chicago.
Note: Shaded areas indicate national recession periods as defined by the National Bureau of Economic Research, Inc. P = peak and T = trough.

CHART K 8
COMMERCIAL BANK LOANS, KALAMAZOO - PORTAGE SMSA
(Current dollars)

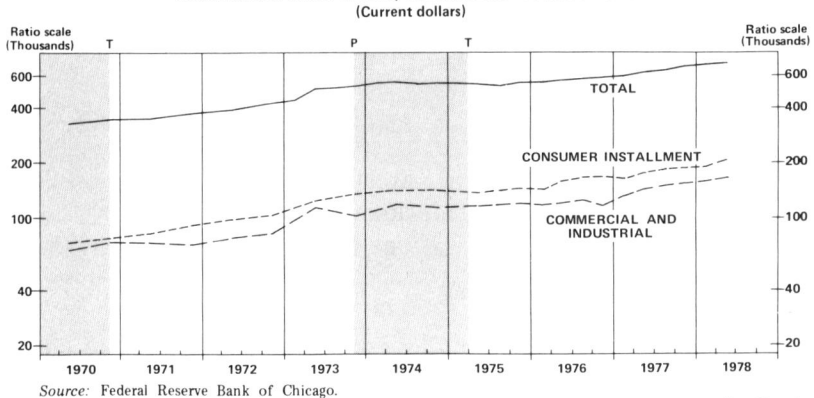

Source: Federal Reserve Bank of Chicago.
Notes: Seasonal adjustment of total loans by the W.E. Upjohn Institute. Other loans are not seasonally adjusted.
Shaded areas indicate national recession periods as defined by the National Bureau of Economic Research, Inc. P = peak and T = trough.

price changes and put in index form, 1972=100. As is the case in nearly all other metropolitan areas in the state, both indexes began to fall before the 1973-75 recession started, and their cyclical declines continued for several quarters after the recession ended. Of course, the cyclical downswing in the total deposits index, 14.7 percent, was less severe than that of the Index of Deflated Demand Deposits which fell 25.1 percent; but while the latter was milder than the decline statewide, the former was more severe. In fact, the downswing in deflated total deposits in the Kalamazoo-Portage SMSA was the fourth most severe among the state's 11 metropolitan areas. Several areas which experienced severe declines in manufacturing employment due to the slump in the automobile industry during the recession, Ann Arbor-Ypsilanti, Flint, and Saginaw, for example, suffered milder setbacks in total deposits than the Kalamazoo-Portage SMSA. Since the recession low, both series have expanded, but by mid-1978 neither had surpassed its pre-recession peak.

The current-dollar volume of loans is shown in Chart K8. Consumer installment loans experienced a slowdown in growth during the recession, but that was rather brief. The growth of commercial and industrial loans also slowed substantially as a result of the recession. But after remaining fairly level from 1974 to 1976, they expanded in 1977 and, like consumer installment loans, were considerably above recession levels as of mid-1978. Total loans in the Kalamazoo-Portage SMSA were more cyclically sensitive, declining 5.1 percent in current dollars during the recession. Only the Battle Creek, Detroit, and Jackson SMSAs had relatively larger declines in total loans. However, as Chart K8 shows, the current-dollar volume of total loans in the area rose substantially during the recent expansion. In the second quarter of 1978, they were up 32.1 percent from a low in 1975 and 25.5 percent above the 1974 peak.

Overall, then, the Kalamazoo-Portage SMSA was the most cyclically stable metropolitan area in the state in terms of wage and salary employment and the unemployment rate. However, indicators of local construction and banking activity exhibited a higher degree of cyclical volatility. Nevertheless, this area ranks with the Grand Rapids SMSA in terms of overall cyclical stability.

Lansing-East Lansing SMSA

The Lansing-East Lansing SMSA includes four counties in the south central part of the lower peninsula: Ionia, Eaton, Clinton, and Ingham. From 1970 to 1977, the population of Eaton County rose 15.6 percent, followed by increases of 12.5 percent in Clinton, 6.8 percent in Ionia, and 4.2 percent in Ingham County. The result was a 7.3 percent increase in the population of the Lansing-East Lansing SMSA over that period, the largest rise among the state's metropolitan areas and considerably above the 2.8 percent increase statewide. In 1977, the Lansing-East Lansing SMSA had an estimated population of 455,100, making this metropolitan area the fourth largest in Michigan with 5 percent of the state total. Although four counties comprise the metropolitan area, 60 percent of the population lives in Ingham County, where the central cities of Lansing and East Lansing are located.

This metropolitan area also ranked fourth in terms of total personal income, estimated at $2,853 million in 1976. Over the 1969-76 period, personal income in the area rose at an annual rate of 9.3 percent, third highest among the 11 metropolitan areas in the state. However, the Lansing-East Lansing SMSA ranked only seventh in terms of per capita income. In 1976, its per capita income was $6,377, which was below the statewide level of $6,757 and slightly less than the nationwide level of $6,396.

Table L1 shows that this metropolitan area, which has the state capital located in the City of Lansing and Michigan State University in East Lansing, has a large government sector compared to the state as a whole. In 1977, 35.2 percent of the area's wage and salary workers were employed in publicly funded jobs, nearly twice the size of the government employment share statewide. Only the Ann Arbor-Ypsilanti SMSA, with 32.6 percent of its wage and salary employment accounted for by government, had a public sector whose relative share came close to that of the Lansing-East Lansing SMSA. The 10 other Michigan metropolitan areas had less than 20 percent of their wage and salary workers employed by the public sector in 1977.

TABLE L1

Percentage Distribution of Total Wage and Salary Employment
Lansing-East Lansing SMSA and Michigan, 1972 and 1977

	Lansing-East Lansing SMSA		Michigan	
Item	1972	1977	1972	1977
Total Wage and Salary	100.0%	100.0%	100.0%	100.0%
Manufacturing	25.0	22.9	35.1	32.4
Durables	22.2	20.2	28.1	25.9
Nondurables	2.8	2.7	7.0	6.5
Nonmanufacturing	42.0	41.9	48.0	49.6
Government	33.0	35.2	16.9	18.0

Source: Michigan Employment Security Commission.

As Table L1 shows, the manufacturing sector accounts for a smaller proportion of wage and salary employment in this area than statewide. In 1977, 22.9 percent of total wage and salary employment in the Lansing-East Lansing SMSA was accounted for by manufacturing industries. And manufacturing in the area was heavily concentrated in durable goods production linked closely to the automobile industry. In 1977, 63.8 percent of the area's manufacturing workers were employed by the transportation equipment industry and another 11.6 percent employed by the metal industry. Among the state's metropolitan areas only the Flint SMSA had a higher percentage of its manufacturing workers employed by the transportation equipment industry. Of course, the Lansing-East Lansing SMSA has a relatively high wage structure and a relatively long workweek, like other areas heavily dependent on the automobile industry. However, it is this industry that largely accounts for the sizable cyclical swings in manufacturing employment in this part of the state.

Business Conditions in the Lansing-East Lansing SMSA

Labor Market Conditions

The major employment categories, shown in Chart L1, reveal a fairly high degree of cyclical sensitivity in local manufacturing employment and relative stability in nonmanufacturing and government employment. The latter was unaffected by the 1973-75 recession and continued to grow. That growth was fairly steady throughout the 1970-78 period at an annual rate of 3.5 percent, slightly above the 3 percent growth statewide (see Table L2). Nonmanufacturing in the area did fall off for a brief period in late 1974, but the decline was very slight. In contrast, employment in the cyclically sensitive manufacturing sector fell 24.5 percent during the recession, the fourth most severe contraction among the state's 11 metropolitan areas. Because that fairly sizable drop directly affected less than one-fourth of the workers in the area, total wage and salary employment experienced a moderate decline of 3.9 percent during the recession. That was the second mildest slump among the metropolitan areas in the state, ranking just behind the 3.2 percent decline in total wage and salary employment in the Kalamazoo-Portage SMSA.

TABLE L2

Average Annual Growth Rates of Selected Labor Market Indicators Lansing-East Lansing SMSA and Michigan[a]

(percent)

Indicator	Lansing-East Lansing SMSA	Michigan
Total wage and salary employment	2.7	1.8
Manufacturing employment	0.6	0.2
Nonmanufacturing employment	3.2	2.6
Government employment	3.5	3.0
Civilian labor force	3.0	2.0
Unemployment rate	3.3	2.8
Average weekly initial claims for UI[b]	4.4	7.3
Average workweek, production workers, mfg.	0.6	0.6

[a] Except where indicated otherwise, estimated growth rates are based on log-linear trends for the 1970-78 period.

[b] Computed for the 1968-78 period.

Chart L1 shows that each employment category has increased substantially during the recent expansion. By the end of 1978, the area's manufacturing employment stood 2.1 percent above its pre-recession peak. As a result of its fairly strong recovery since the recession, manufacturing employment exhibited a slight positive growth for the nine-year period of 0.6 percent per year. Table L2 shows that for the 1970-78 period each of the four employment categories grew more rapidly in the Lansing-East Lansing SMSA than in the state as a whole. The 2.7 percent annual growth rate in total wage and salary employment ranks third, along with the Kalamazoo-Portage SMSA, among the 11 metropolitan areas in Michigan, behind the 3.1 percent and 2.8 percent growth in the Grand Rapids and Ann Arbor-Ypsilanti SMSAs, respectively.

Chart L2 shows that the jobless rate in the Lansing-East Lansing SMSA moved closely with the state average during the 1970-78 period. The local unemployment rate generally remained below the state average during the expansion periods from 1971 to 1973 and 1975 to 1978, but was about equal to the statewide rate during recessions. During the 1973-75 slump, the jobless rate in the Lansing-East Lansing SMSA rose from 4.4 percent to 13.4 percent, compared to the rise in the statewide rate from 5.4 percent to 13.4 percent. By the last quarter of 1978 the local unemployment rate had fallen to 6.0 percent, a full percentage point below the state quarterly average. However, over the nine-year period the unemployment rate in the Lansing-East Lansing SMSA drifted upward at a 3.3 percent annual rate, compared to 2.8 percent for the state as a whole. That represents the fifth highest growth rate among Michigan metropolitan areas.

Average weekly initial claims for unemployment insurance exhibited a pattern of behavior over the 1968-78 period quite similar to the state as a whole (see Chart L3). The major difference between the two is that the local series jumped up considerably at the beginning of each recession, most notably at the start of the 1973-75 slump. Similar behavior occurred in the Flint SMSA, whose manufacturing sector is also dominated by General Motors.

Average weekly initial claims have fallen substantially since the recession but, as Chart L3 shows, the levels in 1978 were still above the cyclical low in 1973. This situation is evident in all Michigan metropolitan areas, the state as a whole, and the nation.[1] However,

---

1. In the Saginaw SMSA, the quarterly average has twice dipped below pre-recession levels but has not remained there.

Lansing-East Lansing SMSA

CHART L 1
WAGE AND SALARY EMPLOYMENT, LANSING-EAST LANSING SMSA
(Seasonally adjusted)

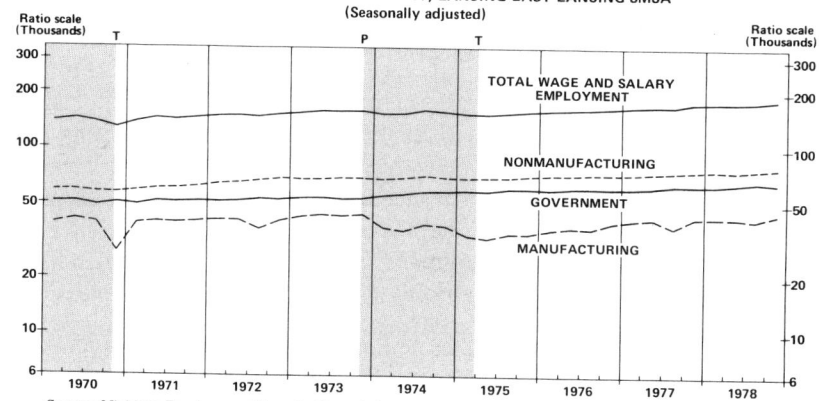

Source: Michigan Employment Security Commission.
Notes: Seasonal adjustment by the W.E. Upjohn Institute.
Shaded areas indicate national recession periods as defined by the National Bureau of Economic Research, Inc.
P = peak and T = trough.

CHART L 2
UNEMPLOYMENT RATE
LANSING - E. LANSING SMSA
(Seasonally adjusted)

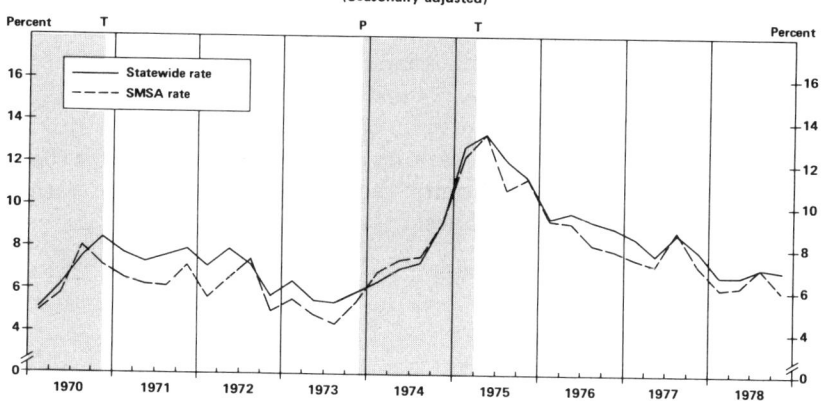

Source: Michigan Employment Security Commission.
Note: Seasonal adjustment by the W.E. Upjohn Institute.
Shaded areas indicate national recession periods as defined by the National Bureau of Economic Research, Inc.
P = peak and T = trough.

Lansing-East Lansing SMSA

CHART L 3
AVERAGE WEEKLY INITIAL CLAIMS FOR UNEMPLOYMENT INSURANCE, STATE PROGRAMS
LANSING-EAST LANSING SMSA
(Seasonally adjusted)

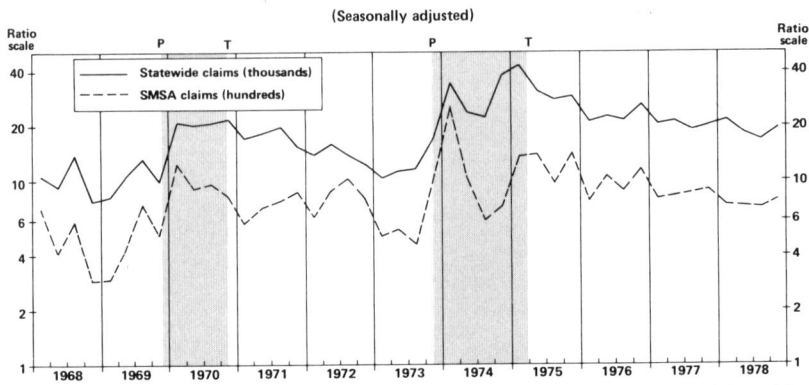

— Statewide claims (thousands)
--- SMSA claims (hundreds)

Source: The W.E. Upjohn Institute, based on data from the Michigan Employment Security Commission and the U.S. Department of Labor.
Shaded areas indicate national recession periods as defined by the National Bureau of Economic Research, Inc.
P = peak and T = trough.

CHART L 4
AVERAGE WEEKLY HOURS OF PRODUCTION WORKERS IN MANUFACTURING INDUSTRIES
LANSING — EAST LANSING SMSA
(Seasonally adjusted)

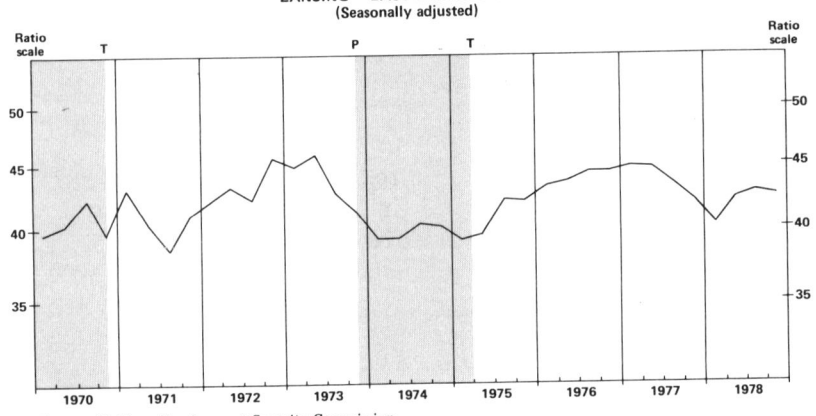

Source: Michigan Employment Security Commission.
Note: Seasonal adjustment by the W.E. Upjohn Institute.
Shaded areas indicate national recession periods as defined by the National Bureau of Economic Research, Inc.
P = peak and T = trough.

for the entire 1968-78 period, initial claims in the Lansing-East Lansing SMSA grew at a slower annual rate than the state, 4.4 percent compared to 7.3 percent, and that local growth rate ranks fifth lowest among the 11 metropolitan areas in Michigan, well below the most rapid growth of 11.1 percent per year in the Ann Arbor-Ypsilanti SMSA.

The average workweek of manufacturing production workers in the Lansing-East Lansing SMSA was the most cyclically volatile among the metropolitan areas in the state. Of course, this is not surprising, given the heavy dependence on automobile production in the area's manufacturing sector. Other metropolitan areas in Michigan with similar extreme concentrations in manufacturing, for example, Flint, Saginaw, and Ann Arbor-Ypsilanti, also exhibit a high degree of volatility. Chart L4 shows that the average workweek rose to almost 46 hours in the second quarter of 1973, then plunged 14.6 percent during the recession. That contraction was the most severe among the state's metropolitan areas, ranking just ahead of the decline in the Flint SMSA. Since the recession, the average workweek in the area has risen substantially, but the most recent high of 44.8 hours in early 1977 was still below the 1973 peak. In both the Flint and Saginaw SMSAs, where the sensitivity of the average workweek is similar to that of the Lansing-East Lansing SMSA, hours worked in manufacturing have risen above pre-recession levels during the current expansion.

Construction

Construction activity in the area is highly cyclically sensitive. Chart L5 shows that new building permits for private housing moved up and down with swings in national business activity, but with a lead particularly at peaks. As the chart shows, the moving average of the unadjusted Index of New Building Permits began falling about seven quarters before the 1969-70 recession started, and the following upswing began before the national recession ended. This behavior is exactly the same as the national Index of New Building Permits during that period. The local downswing was much more severe, however. From peak to trough the moving average fell 63.4 percent in this SMSA, compared to a 24.5 percent drop in the national series and a 21.6 percent drop statewide. That slump in new building permits in the Lansing-East Lansing SMSA was the second most severe among the state's 11 metropolitan areas.

Like most other areas in Michigan, new building permits recovered strongly during the 1970-73 expansion, then plunged 71.2

154　Lansing-East Lansing SMSA

CHART L 5
INDEX OF NEW BUILDING PERMITS, PRIVATE HOUSING
LANSING - EAST LANSING SMSA
(1967 = 100)

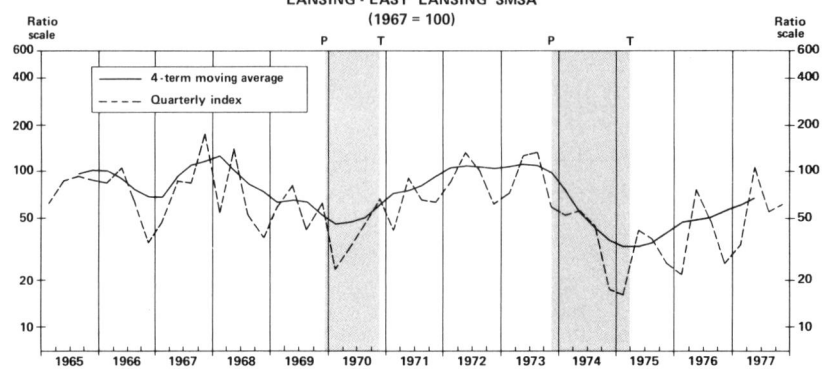

Source: The W.E. Upjohn Institute. Index based on the U.S. Department of Commerce, Bureau of the Census, *Construction Reports — Housing Authorized by Building Permits and Public Contracts, C-40.*
Shaded areas indicate national recession periods as defined by the National Bureau of Economic Research, Inc. P = peak and T = trough.

CHART L 6
CONSTRUCTION EMPLOYMENT, LANSING-EAST LANSING SMSA
(Seasonally adjusted)

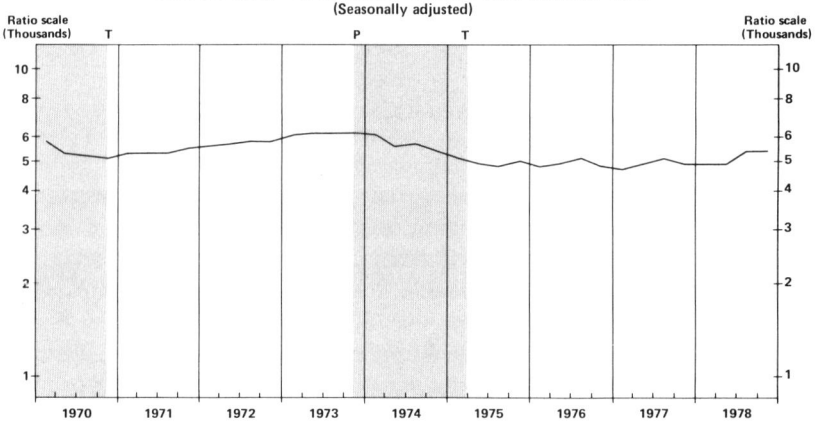

Source: Michigan Employment Security Commission.
Notes: Seasonal adjustment by the W.E. Upjohn Institute.
Shaded areas indicate national recession periods as defined by the National Bureau of Economic Research, Inc. P = peak and T = trough.

percent during the 1973-75 recession. Since the low in early 1975, new building permits in the area have risen, but preliminary data for 1978 show that recent highs in the moving average are still considerably below the pre-recession peak. A similar situation exists in the state, the nation, and all Michigan metropolitan areas except the Jackson SMSA.

Employment in the local construction industry was especially hard hit during the last recession, dropping 24.2 percent from a high in the second quarter of 1973 to a low in the first quarter of 1977. That represents one of the longest slumps in construction employment among Michigan metropolitan areas. Chart L6 shows that construction employment in the area remains depressed, with little upward momentum evident. As a result, the Lansing-East Lansing SMSA has had the weakest recovery for this series among all Michigan metropolitan areas. By the last quarter of 1978, local construction employment was still 13 percent below its pre-recession level. Because of this weak recovery following a relatively severe slump, construction employment declined at an annual rate of 1.5 percent over the entire 1970-78 period. The Lansing-East Lansing SMSA, therefore, joins five other metropolitan areas in the state, Ann Arbor-Ypsilanti, Battle Creek, Flint, Jackson, and Saginaw, where employment in the construction industry exhibited negative growth over the nine-year period.

Banking Activity

Deposits and loans at commercial banks performed somewhat more poorly in the Lansing-East Lansing SMSA than in the state as a whole over the 1970-78 period. Table L3 shows that the growth in total loans and deposits was slower in the local area than statewide. Moreover, the 6.6 percent annual growth in the current-dollar volume of total loans was second lowest among the metropolitan areas in Michigan, ranking just above the 6.3 percent growth in the Detroit SMSA. However, the 10.2 percent annual growth in consumer installment loans ranked fourth highest and commercial and industrial loans fifth highest among Michigan metropolitan areas, and both rates exceeded the growth in those loan categories in the state as a whole.

Chart L7 shows that after adjustment for price increases, demand and total deposits in the area experienced a long downswing lasting 18 and 13 quarters, respectively. Those declines were also rather sizable. Deflated demand deposits fell 28.6 percent, and deflated total deposits declined 15.7 percent over that period. The former was the fourth most severe drop among the 11 metropolitan areas and the latter ranked second, along with the decline in the Battle Creek SMSA. In addition, Chart L7 reveals that expansions in both local series have been rather weak. As a result, in mid-1978 both indexes were still well below pre-recession levels.

TABLE L3

Average Annual Growth Rates of
Selected Commercial Banking Indicators
Lansing-East Lansing SMSA and Michigan[a]

(percent)

Indicator	Lansing-East Lansing SMSA	Michigan
Demand deposits (current dollars)	3.9	3.7
Deflated demand deposits[b]	-2.8	-3.0
Total deposits (current dollars)	6.2	7.3
Deflated total deposits[b]	-0.6	0.4
Total loans (current dollars)	6.6	7.6
Commercial and industrial loans (current dollars)	8.7	7.3
Consumer installment loans (current dollars)	10.2	9.1

[a]Except where indicated otherwise, estimated growth rates are based on log-linear trends for the 1970-78 period.

[b]Current-dollar values adjusted for changes in the U.S. Consumer Price Index.

The current-dollar volume of total loans fell briefly during the last recession by about 5.1 percent. That slump ranks fourth in severity among the state's metropolitan areas. Brief declines were also evident in consumer installment loans and commercial and industrial loans during the recession. Chart L8 shows, however, that the current-dollar volume of loans has expanded considerably since the recession. By mid-1978 total loans had increased 17.9 percent from the cyclical low in the second quarter of 1975.

In summary, the Lansing-East Lansing SMSA exhibited a good deal of variation in growth and fluctuations among local economic activities. Total wage and salary employment was characterized by relative cyclical stability and fairly strong growth compared to

Lansing-East Lansing SMSA 157

CHART L 7
INDEX OF DEFLATED TOTAL DEPOSITS AND INDEX OF DEFLATED DEMAND DEPOSITS
LANSING - EAST LANSING SMSA
(1972 = 100)
(Seasonally adjusted)

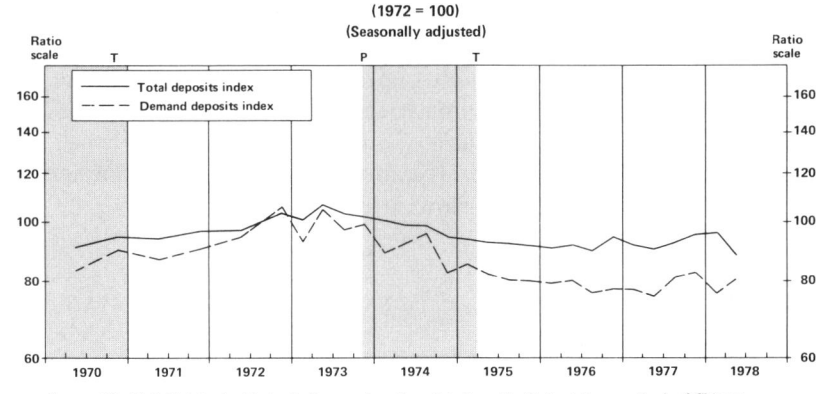

Source: The W.E. Upjohn Institute. Indices are based on data from the Federal Reserve Bank of Chicago.
Note: Shaded areas indicate national recession periods as defined by the National Bureau of Economic Research, Inc. P = peak and T = trough.

CHART L 8
COMMERCIAL BANK LOANS, LANSING - EAST LANSING SMSA
(Current dollars)

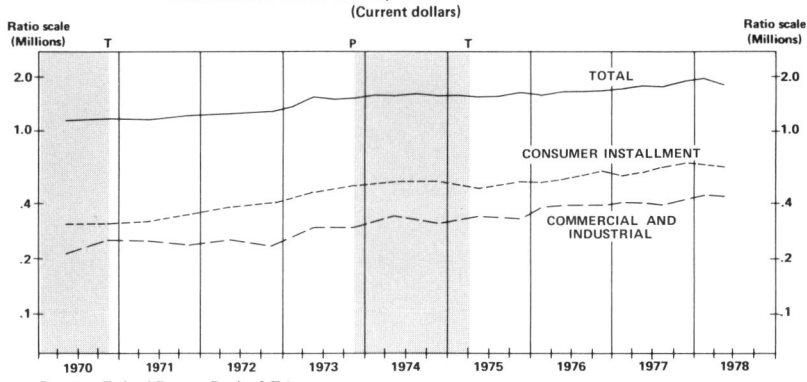

Source: Federal Reserve Bank of Chicago.
Notes: Seasonal adjustment of total loans by the W.E. Upjohn Institute. Other loans are not seasonally adjusted. Shaded areas indicate national recession periods as defined by the National Bureau of Economic Research, Inc. P = peak and T = trough.

other metropolitan areas in the state. On the other hand, the manufacturing sector exhibited a good deal of cyclical volatility, with employment and hours worked experiencing sizable upswings and downswings. Construction activity was also subject to large cyclical swings. A most notable characteristic of this area was the relatively weak recovery in construction employment after the last recession. Finally, the area's banking performance was not particularly strong, with total loans and deposits at commercial banks growing at relatively slow rates.

Muskegon-Norton Shores-Muskegon Heights SMSA

The Muskegon-Norton Shores-Muskegon Heights SMSA (hereafter referred to as the Muskegon SMSA) is made up of two counties, Muskegon and Oceana, located in the west central part of the lower peninsula on the Lake Michigan shoreline. In terms of population, it is one of the smallest metropolitan areas in Michigan. In 1977, its population of 179,000 made it the third smallest metropolitan area, ranking just ahead of the Bay City and Jackson SMSAs. Between 1970 and 1977, this area's population rose by only 2.1 percent, with the population of Muskegon County rising a mere 0.4 percent. During this eight-year period the two-county region experienced an estimated net outmigration of 3.4 percent, which offset a natural rise—births exceeding deaths—of 5.5 percent,resulting in a modest overall population increase. Among the state's 11 metropolitan areas, only Battle Creek, Detroit, and Flint experienced slower population growth.

Total personal income in the area amounted to $980 million in 1976, ninth largest among the 11 metropolitan areas. However, the Muskegon SMSA has the distinction of having the lowest per capita income. In 1976, the area's per capita income was $5,496, which was well below the statewide level of $6,757 and the high of $7,496 in the Detroit SMSA. In the same year, average weekly manufacturing wages in the Muskegon SMSA were the second lowest in the state. Moreover, income growth was quite slow during the 1969-76 period, with personal income rising at an annual rate of 7.7 percent, second lowest among the state's metropolitan areas.

Table M1 shows that the Muskegon SMSA is more dependent on the manufacturing sector for jobs than is the state as a whole. In 1972, 42.1 percent of its wage and salary workers were employed by manufacturing firms. That proportion had fallen to 37.0 percent by

1977. Only the Saginaw and Flint SMSAs are more dependent on manufacturing, each having more than 40 percent of their wage and salary workers employed by manufacturing firms.

TABLE M1

Percentage Distribution of Total Wage and Salary Employment
Muskegon-Norton Shores-Muskegon Heights SMSA and Michigan
1972 and 1977

| | Muskegon SMSA | | Michigan | |
Item	1972	1977	1972	1977
Total Wage and Salary.....	100.0%	100.0%	100.0%	100.0%
Manufacturing	42.1	37.0	35.1	32.4
Durables.............	36.0	31.6	28.1	25.9
Nondurables	6.1	5.4	7.0	6.5
Nonmanufacturing	42.8	45.4	48.0	49.6
Government	15.1	17.6	16.9	18.0

Source: Michigan Employment Security Commission.

Table M1 also shows that this area's manufacturing employment is heavily concentrated in durable goods industries. The metal and nonelectrical machinery industries are the major employers in the area, accounting for slightly more than 68 percent of manufacturing employment in 1977. However, those industries have not been expanding in the Muskegon SMSA and, as a result, the area's manufacturing base has been shrinking for some time. Thus, the area's economy exhibited behavior characteristic of declining urban areas: slow growth of population, employment and income, and a high unemployment rate over a relatively long period of time.

Business Conditions in the Muskegon SMSA

Labor Market Conditions

The most notable feature of total wage and salary employment in this area during the 1970-78 period is its relatively slow growth. Table M2 shows that wage and salary employment grew at an annual rate of 1.5 percent over that nine-year period, compared to 1.8

percent for the state as a whole. Moreover, the employment growth in the Muskegon SMSA only ranks ahead of the three slowest growth areas in Michigan, Battle Creek, Jackson, and Detroit, which had annual growth rates for wage and salary employment of 0.9 percent, 1.1 percent, and 1.3 percent, respectively, over that nine-year period. In addition, the growth of the labor force in the area was quite modest, 1.4 percent per year. Only the Battle Creek and Detroit SMSAs exhibited slower labor force growth from 1970 to 1978.

TABLE M2

Average Annual Growth Rates of
Selected Labor Market Indicators
Muskegon-Norton Shores-Muskegon Heights SMSA and Michigan[a]

(percent)

Indicator	Muskegon Heights SMSA	Michigan
Total wage and salary employment	1.5	1.8
Manufacturing employment	-1.1	0.2
Nonmanufacturing employment	3.0	2.6
Government employment	4.5	3.0
Civilian labor force	1.4	2.0
Unemployment rate	1.0	2.8
Average weekly initial claims for UI[b]	6.8	7.3
Average workweek, production workers, mfg.	0.4	0.6

[a] Except where indicated otherwise, estimated growth rates are based on log-linear trends for the 1970-78 period.

[b] Computed for the 1968-78 period.

Like the Battle Creek, Detroit, and Jackson SMSAs, the Muskegon SMSA experienced negative growth in manufacturing employment over the 1970-78 period. It is the decline in the area's manufacturing base, reflected in this downtrend, that accounts for much of the slow growth in total wage and salary employment. As Chart M1 shows, the long-term decline in manufacturing reduced that sector from the largest in the area to the second largest, behind nonmanufacturing industries, a situation that has existed since the last recession. During the recession, local manufacturing employ-

162 Muskegon-Norton Shores-Muskegon Heights SMSA

CHART M 1
WAGE AND SALARY EMPLOYMENT, MUSKEGON-NORTON SHORES-MUSKEGON HEIGHTS SMSA
(Seasonally adjusted)

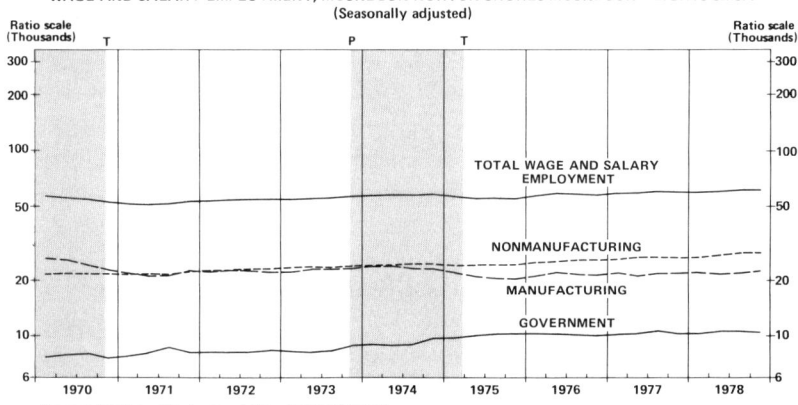

Source: Michigan Employment Security Commission.
Notes: Seasonal adjustment by the W.E. Upjohn Institute.
Shaded areas indicate national recession periods as defined by the National Bureau of Economic Research, Inc
P = peak and T = trough.

CHART M 2
UNEMPLOYMENT RATE
MUSKEGON - NORTON SHORES - MUSKEGON HEIGHTS SMSA
(Seasonally adjusted)

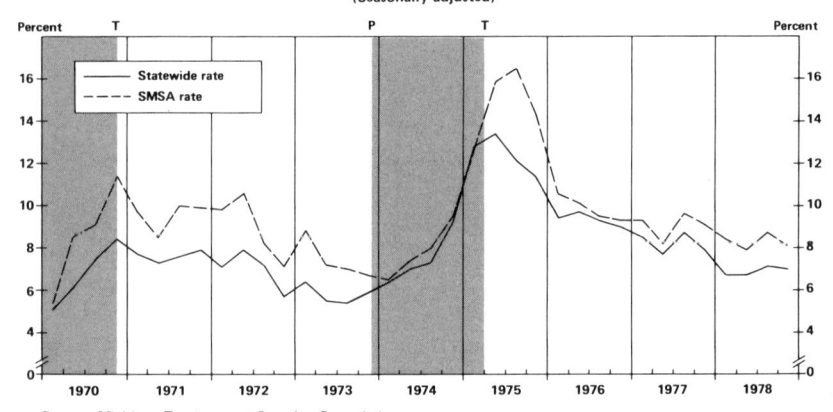

Source: Michigan Employment Security Commission.
Note: Seasonal adjustment by the W.E. Upjohn Institute.
Shaded areas indicate national recession periods as defined by the National Bureau of Economic Research, Inc.
P = peak and T = trough.

ment fell 15.3 percent. While that was the second mildest cyclical decline among the state's 11 metropolitan areas, it was followed by a weak recovery in which, as Chart M1 shows, the area's manufacturing employment remained below its pre-recession peak. By the end of 1978, local manufacturing employment was 11.7 percent above its recession low of late 1975, but still 5.4 percent below its 1974 high. Only in the Battle Creek and Jackson SMSAs were manufacturing employment levels in the last quarter of 1978 farther below their pre-recession peaks.

Because of the relative stability of nonmanufacturing employment and the rise in government employment in the area during the last recession, total wage and salary employment behaved comparatively well, declining just 4.3 percent from peak to trough. That was the third mildest contraction among the state's metropolitan areas, ranking just behind the declines of 3.2 percent and 3.9 percent in the Kalamazoo-Portage and Lansing-East Lansing SMSAs, respectively. However, unlike those two areas, the expansion of wage and salary employment in the Muskegon SMSA since the recession has been quite modest. By the fourth quarter of 1978, this area's total wage and salary employment was 6 percent above its 1974 high. Only the Battle Creek, Detroit, and Jackson SMSAs performed more poorly in this respect.

The weakness of the local economy is also reflected in its unemployment rate which, as Chart M2 shows, remained above the state average throughout the 1970-78 period. The local jobless rate rose to a seasonally-adjusted high of 11.4 percent in the fourth quarter of 1970. Among the state's 11 metropolitan areas, only Flint registered a higher jobless rate during the 1969-70 recession. During the expansion period from 1971 through 1973, Muskegon was the only metropolitan area in Michigan where the unemployment rate remained above 6 percent, with the cyclical low of 6.5 percent occurring in the first quarter of 1974. The 1973-75 national recession, combined with a shrinking manufacturing base in the area, resulted in a 16.5 percent unemployment rate by the third quarter of 1975. Again, only the Flint SMSA registered a higher unemployment rate. Chart M2 shows that while the local unemployment rate has fallen considerably during the recent business expansion, it remained over 8 percent for 1978, highest among Michigan metropolitan areas and well above the state average. It is quite clear from the behavior of the jobless rate that the unemployment situation was worse in this metropolitan area than in any other in Michigan over the 1970-78 period.

164   Muskegon-Norton Shores-Muskegon Heights SMSA

CHART M 3
AVERAGE WEEKLY INITIAL CLAIMS FOR UNEMPLOYMENT INSURANCE, STATE PROGRAMS
MUSKEGON, NORTON SHORES, MUSKEGON HEIGHTS SMSA
(Seasonally adjusted)

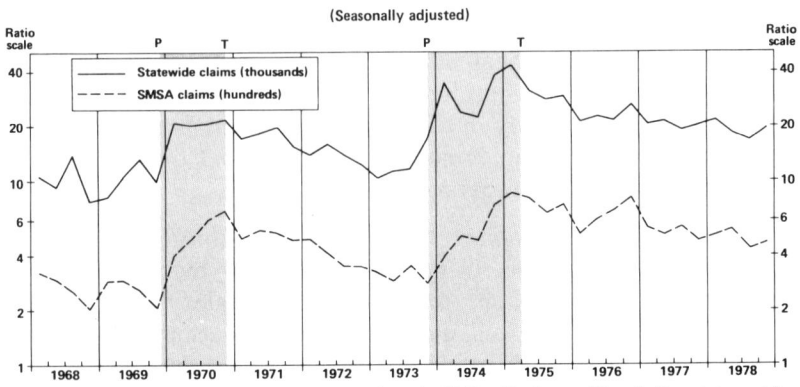

Source: The W.E. Upjohn Institute, based on data from the Michigan Employment Security Commission and the U.S. Department of Labor.
Shaded areas indicate national recession periods as defined by the National Bureau of Economic Research, Inc.
P = peak and T = trough.

CHART M 4
AVERAGE WEEKLY HOURS OF PRODUCTION WORKERS IN MANUFACTURING INDUSTRIES
MUSKEGON – NORTON SHORES – MUSKEGON HEIGHTS SMSA
(Seasonally adjusted)

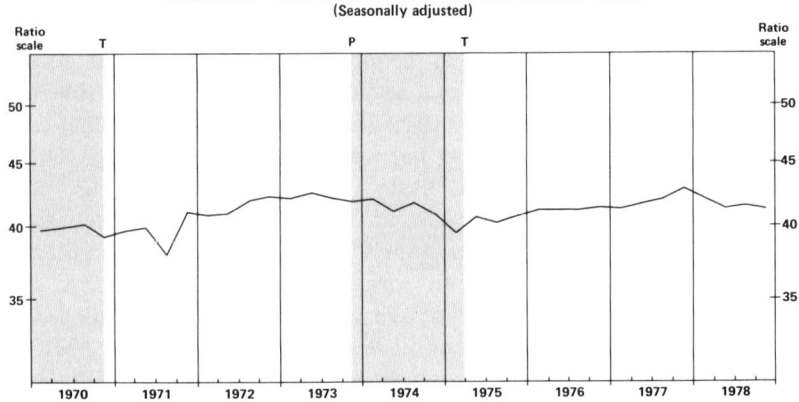

Source: Michigan Employment Security Commission.
Note: Seasonal adjustment by the W.E. Upjohn Institute.
Shaded areas indicate national recession periods as defined by the National Bureau of Economic Research, Inc.
P = peak and T = trough.

There is nothing particularly surprising about the behavior of average weekly initial claims for unemployment insurance in the Muskegon SMSA. Chart M3 shows that over the 1968-78 period, state and local claims tended to rise and fall in similar cyclical patterns. During the 1969-70 recession, initial claims rose 227.2 percent in the Muskegon SMSA, compared to 193.2 percent statewide. However, the magnitude of the cyclical upswing during the 1973-75 recession was greater for the state as a whole than for the local area, 301.1 percent compared to 198.3 percent. The declines in initial claims during the business expansion periods were about the same. Table M2 shows that for the entire eleven-year period, average weekly initial claims for unemployment insurance grew at an annual rate of 6.8 percent, slightly below the statewide growth rate of 7.3 percent. The growth rate in the Muskegon SMSA ranked below that of four other Michigan metropolitan areas, an interesting feature of local labor market behavior which may be due to the relatively slow growth of employment and the high unemployment rate in this area, which did not allow insured UI claimants to requalify for benefits.

Finally, the average workweek of manufacturing production workers varied less in the Muskegon SMSA than in the state as a whole. Chart M4 shows that the average workweek rose to about 42.5 hours in the second quarter of 1973, below the peak of 44.2 hours for the state as a whole. The decline of 7.1 percent during the 1973-75 recession was less severe than the statewide contraction of 9.7 percent. In addition, six other metropolitan areas in Michigan experienced more severe declines during the last business slump. The average workweek in the area has recovered substantially, rising in late 1977 to a level above the pre-recession peak. That situation has not occurred for the state as whole.

Construction

Building activity in the Muskegon SMSA did not reflect the weaknesses evident in the local labor market. Both new building permits for private housing and employment in the local construction industry exhibited a fair amount of strength compared to many other metropolitan areas in Michigan. This is somewhat surprising given the relatively weak performance of income and employment in this area.

Chart M5 shows the Index of New Building Permits by quarter in its unadjusted form (dashed line) and a moving average (solid line). As is the case in other Michigan metropolitan areas, new building permits in the Muskegon SMSA move in a cyclical manner, generally conforming to national business cycle patterns, with a lead at peaks. The magnitude of the cyclical swings was larger for this local area compared to the state and nation. During the 1969-70 recession, new building permits (the moving average in Chart M5) fell 33.1 percent in the Muskegon SMSA. The decline statewide was a comparatively modest 21.6 percent, while nationwide new building permits fell 24.5 percent. Of course, the contraction during the 1973-75 national recession was considerably more severe. New building permits dropped 72.7 percent in the Muskegon SMSA, 56.2 percent in the state as a whole, and 69.1 percent nationwide. It should be noted that during this period, 9 of 11 metropolitan areas in Michigan experienced a more sizable falloff in new building permits than did the state as a whole. Thus, the behavior in the Muskegon SMSA is quite similar to most other metropolitan areas in the state during the recession in the mid-1970s.

Chart M5 shows that the expansion of new building permits since the recession appears to be rather short, with a decline evident from mid-1976 to mid-1977. Preliminary data for 1978 reveal, however, that the Index of New Building Permits in the area again moved upward after a good deal of sluggishness in 1977. The most recent quarterly data available put the moving average above the high in 1976, but still about 19 percent below the pre-recession peak. A similar situation exists in all Michigan metropolitan areas except Jackson, where new building permits moved above pre-recession levels during the recent expansion period.

Employment in the local construction industry, shown in Chart M6, was relatively strong throughout the 1970-78 period. The most notable characteristic is its strong upward trend. Over the nine-year period construction employment grew at an annual rate of 5.1 percent, the highest growth rate among Michigan metropolitan areas. This fairly strong growth is even more outstanding when it is realized that 6 of 11 metropolitan areas in the state experienced negative growth over those nine years. Also, the cyclical decline in construction employment during the 1973-75 recession was comparatively short and mild. From peak to trough, construction employment fell 13.0 percent, the least severe decline among the 11 metropolitan areas in the state and well below the 23.5 percent drop statewide. As Chart M6 shows, construction employment has expanded considerably in this area since the recession. At the end of 1978 it was 30.4 percent above its previous cyclical high in 1974. In

Muskegon-Norton Shores-Muskegon Heights 167

CHART M 5
INDEX OF NEW BUILDING PERMITS, PRIVATE HOUSING
MUSKEGON - NORTON SHORES - MUSKEGON HEIGHTS SMSA
(1967 = 100)

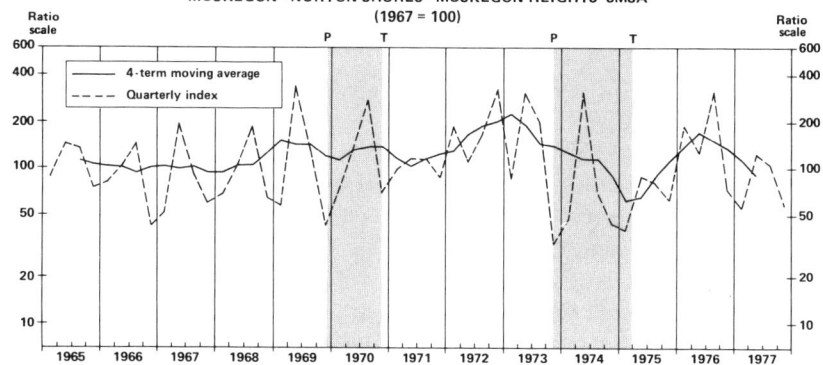

Source: The W.E. Upjohn Institute. Index based on the U.S. Department of Commerce, Bureau of the Census, *Construction Reports — Housing Authorized by Building Permits and Public Contracts, C-40.*
Shaded areas indicate national recession periods as defined by the National Bureau of Economic Research, Inc.
P = peak and T = trough.

CHART M 6
CONSTRUCTION EMPLOYMENT, MUSKEGON-NORTON SHORES-MUSKEGON HEIGHTS SMSA
(Seasonally adjusted)

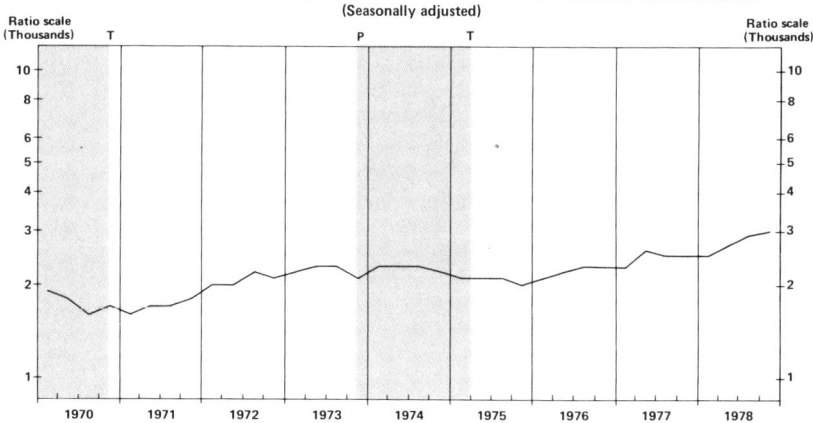

Source: Michigan Employment Security Commission.
Notes: Seasonal adjustment by the W.E. Upjohn Institute.
Shaded areas indicate national recession periods as defined by the National Bureau of Economic Research, Inc.
P = peak and T = trough.

168   Muskegon-Norton Shores-Muskegon Heights SMSA

no other metropolitan area has construction employment exhibited such strength.

Banking Activity

Over the 1970-77 period, loans and deposits at commercial banks grew more rapidly in the Muskegon SMSA than in the state as a whole. Especially impressive is the 9.3 percent annual growth in the current-dollar volume of total deposits which, as Table M3 shows, exceeded the 7.3 percent growth statewide. In addition, the growth of total deposits in the Muskegon SMSA was the third highest of the Michigan metropolitan areas over this eight-year period. The growth in the current-dollar volume of total loans was also relatively rapid in this area, increasing 10.3 percent per year compared to 7.6 percent for the state as a whole. Among the state's 11 metropolitan areas, this growth in total loans ranks second, just behind the 11.6 percent growth rate in the Saginaw SMSA. This is also the case for the 13.3 percent growth in this area's commercial

TABLE M3

Average Annual Growth Rates of
Selected Commercial Banking Indicators
Muskegon-Norton Shores-Muskegon Heights SMSA and Michigan[a]

(percent)

Indicator	Muskegon SMSA	Michigan
Demand deposits (current dollars)	4.0	3.7
Deflated demand deposits[b]	-2.7	-3.0
Total deposits (current dollars)	9.3	7.3
Deflated total deposits[b]	2.4	0.4
Total loans (current dollars)	10.3	7.6
Commercial and industrial loans (current dollars)	13.3	7.3
Consumer installment loans (current dollars)	10.8	9.1

[a] Except where indicated otherwise, estimated growth rates are based on log-linear trends for the 1970-78 period.

[b] Current-dollar values adjusted for changes in the U.S. Consumer Price Index.

and industrial loan volume, which matched the growth rate in the Ann Arbor-Ypsilanti area. The 10.8 percent annual growth in consumer installment loans in the Muskegon SMSA is also relatively rapid. Only the Saginaw and Kalamazoo-Portage SMSAs experienced more rapid growth, 14.5 percent and 12.6 percent, respectively.

Chart M7 shows that, after adjusting for price increases, demand deposits experienced a longer and deeper slump than total deposits during the mid-1970s. This is similar to the behavior of these two indexes in all other Michigan metropolitan areas. The cyclical decline in the Index of Deflated Total Deposits was 11 percent, the fifth mildest downswing among the 11 metropolitan areas in the state. As the chart shows, the total deposits index has recovered substantially during the current business expansion, but by mid-1978 it was still 5.9 percent below its pre-recession peak.

The current-dollar volume of total loans, commercial and industrial loans, and consumer installment loans is shown in Chart M8. Of the three, only consumer installment loans fell during the 1973-75 recession, and that decline was about 6.3 percent from peak to trough. Total loans and commercial and industrial loans did experience a slowdown in their growth rates during the recession but, as Chart M8 shows, this was brief. After that, both continued their fairly rapid expansion, which resulted in levels in 1978 considerably above pre-recession levels. Although consumer installment loans did fall off during the recession, they too experienced a fairly vigorous expansion, so that by mid-1978 the current-dollar volume was 31.0 percent above the previous cyclical high of the fourth quarter of 1973.

In summary, therefore, indicators of economic activity in the Muskegon SMSA present a rather mixed picture of the local economy over roughly the last ten years. There is no question that the manufacturing base has been shrinking and, as a result, employment growth has been weak compared to many other Michigan metropolitan areas. In addition, relatively high unemployment rates exist in this two-county region. However, building activity is not depressed in this area, and construction employment grew more rapidly than in any other metropolitan area in the state during the 1970-78 period. Also, loan activity at local commercial banks has been fairly strong, with no long or severe setbacks evident during the last recession.

170 Muskegon-Norton Shores-Muskegon Heights

CHART M 7
INDEX OF DEFLATED TOTAL DEPOSITS AND INDEX OF DEFLATED DEMAND DEPOSITS
MUSKEGON, NORTON SHORES, MUSKEGON HEIGHTS SMSA
(1972 = 100)
(Seasonally adjusted)

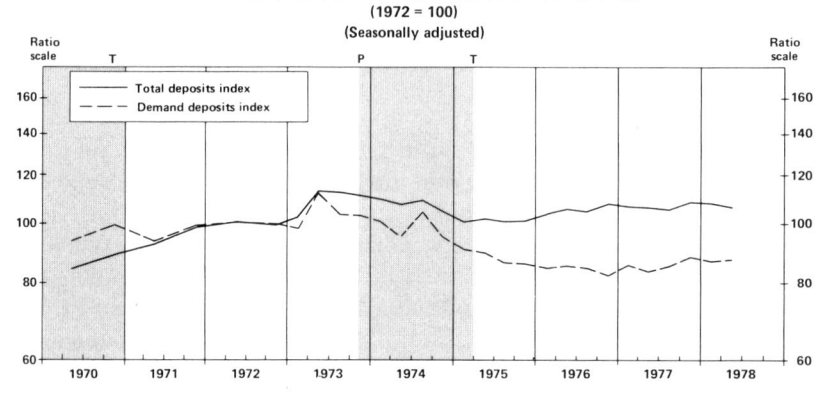

Source: The W.E. Upjohn Institute. Indices are based on data from the Federal Reserve Bank of Chicago.
Note: Shaded areas indicate national recession periods as defined by the National Bureau of Economic Research, Inc. P = peak and T = trough.

CHART M 8
COMMERCIAL BANK LOANS, MUSKEGON - NORTON SHORES - MUSKEGON HEIGHTS SMSA
(Current dollars)

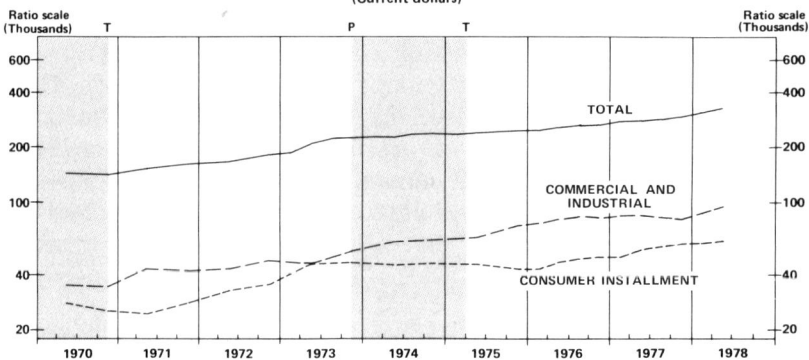

Source: Federal Reserve Bank of Chicago.
Notes: Seasonal adjustment of total loans by the W.E. Upjohn Institute. Other loans are not seasonally adjusted.
Shaded areas indicate national recession periods as defined by the National Bureau of Economic Research, Inc.
P = peak and T = trough.

Saginaw SMSA

The Saginaw SMSA is a single-county metropolitan area—Saginaw County—located in the east central part of the lower peninsula. It is bordered by the Bay City SMSA to the north and the Flint SMSA to the south. In 1977, the population of this area was 226,700, making it the seventh largest among the state's 11 metropolitan areas. From 1970 to 1977 the area's population increased by 3.1 percent, which is slightly above the statewide rise of 2.8 percent.

The Saginaw SMSA also ranks seventh in terms of personal income. In 1976, the area's total personal income amounted to $1,513 million. Over the 1969-76 period, personal income grew at an annual rate of 9.7 percent, second only to the 9.8 percent growth in the Ann Arbor-Ypsilanti SMSA. In terms of per capita income, the Saginaw SMSA ranked fourth among the state's 11 metropolitan areas, with a level of $6,692 in 1976. Average hourly earnings of manufacturing workers in Saginaw are the highest in the state due to the heavy dependence on the automobile industry. As in the Flint SMSA to the south, General Motors is the largest employer in the Saginaw SMSA, giving those two adjacent metropolitan areas the highest hourly-wage rates in Michigan.

Table S1 shows that the Saginaw SMSA is heavily concentrated in manufacturing. In 1977, 41.2 percent of the area's wage and salary workers were employed by manufacturing industries. Only the Flint SMSA had a higher proportion of workers in the manufacturing sector. Also, as Table S1 shows, manufacturing employment is predominantly in durable goods industries. In 1977, 37.8 percent of the area's wage and salary workers were engaged in the production of durable goods, compared to 25.9 percent statewide. The transportation equipment industry accounted for the largest share of manufacturing workers, 41.6 percent, followed by the metal and the nonelectrical machinery industries. Therefore, like the Flint SMSA, this area has an industrial structure dominated by the cyclically sensitive, but high-wage, automobile industry.

TABLE S1

Percentage Distribution of Total Wage and Salary Employment
Saginaw SMSA and Michigan, 1972 and 1977

Item	Saginaw SMSA		Michigan	
	1972	1977	1972	1977
Total Wage and Salary	100.0%	100.0%	100.0%	100.0%
Manufacturing	42.2	41.2	35.1	32.4
Durables	38.8	37.8	28.1	25.9
Nondurables	3.4	3.4	7.0	6.5
Nonmanufacturing	45.9	45.4	48.0	49.6
Government	11.9	13.4	16.9	18.0

Source: Michigan Employment Security Commission.

Business Conditions in the Saginaw SMSA

Labor Market Conditions

Over the 1970-78 period, the Saginaw SMSA experienced more rapid growth and larger cyclical swings in employment than the state as a whole. Table S2 shows that total wage and salary employment in this area grew at an annual rate of 2.4 percent, compared to a statewide growth rate of 1.8 percent. This difference is attributable mainly to the higher growth rates in manufacturing and government employment in the local area than in the state as a whole. The 1.8 percent annual increase in manufacturing employment in the Saginaw SMSA over this nine-year period represents one of the strongest performances among Michigan metropolitan areas. In fact, only the Ann Arbor-Ypsilanti SMSA, with an annual growth rate of 2.1 percent for manufacturing employment, outperformed the Saginaw SMSA in this respect. In addition, the 4.4 percent annual growth in government employment was the third highest among the state's 11 metropolitan areas. Therefore, as Chart S1 shows, this area's total wage and salary employment was bolstered by moderate, but fairly steady, growth of nonmanufacturing employment, combined with relatively rapid growth of employment in local manufacturing industries and in the government sector.

However, Chart S1 reveals that growth in manufacturing employment was not steady throughout the period, but was subject to sizable cyclical swings. Of course this is not surprising, given the area's heavy dependence on the automobile industry. During the 1973-75 recession, local manufacturing employment fell 20.6 percent, exceeding the statewide decline of 18.7 percent and contributing substantially to the 7.7 percent drop in the area's total wage and salary employment. It should be noted, however, that the downswing in manufacturing employment in the Saginaw SMSA ranks as the fifth most severe, behind declines in the Ann Arbor-Ypsilanti, Flint, Jackson, and Lansing-East Lansing SMSAs.

TABLE S2

Average Annual Growth Rates of Selected Labor Market Indicators
Saginaw SMSA and Michigan[a]

(percent)

Indicator	Saginaw SMSA	Michigan
Total wage and salary employment	2.4	1.8
Manufacturing employment	1.8	0.2
Nonmanufacturing employment	2.5	2.6
Government employment	4.4	3.0
Civilian labor force	2.4	2.0
Unemployment rate	3.9	2.8
Average weekly initial claims for UI[b]	3.8	7.3
Average workweek, production workers, mfg.[b]	0.7	0.2

[a] Except where indicated otherwise, estimated growth rates are based on log-linear trends for the 1970-78 period.

[b] Computed for the 1968-78 period.

Chart S1 also shows that while nonmanufacturing employment did fall off during the recession, the decline was brief and fairly mild.

Saginaw SMSA

CHART S 1
WAGE AND SALARY EMPLOYMENT, SAGINAW SMSA
(Seasonally adjusted)

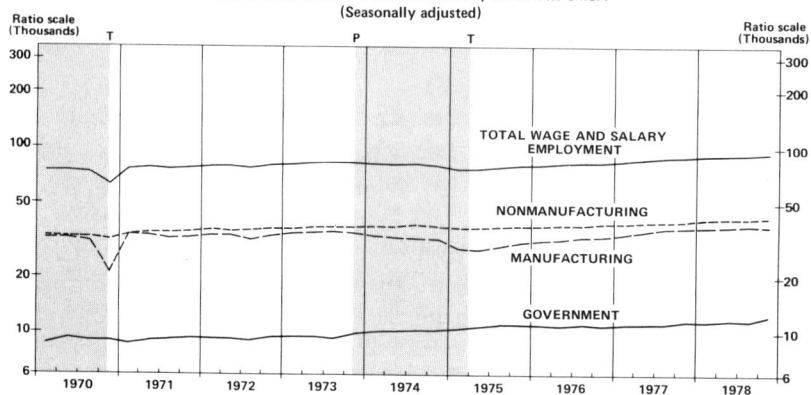

Source: Michigan Employment Security Commission.
Notes: Seasonal adjustment by the W.E. Upjohn Institute.
Shaded areas indicate national recession periods as defined by the National Bureau of Economic Research, Inc.
P = peak and T = trough.

CHART S 2
UNEMPLOYMENT RATE
SAGINAW SMSA
(Seasonally adjusted)

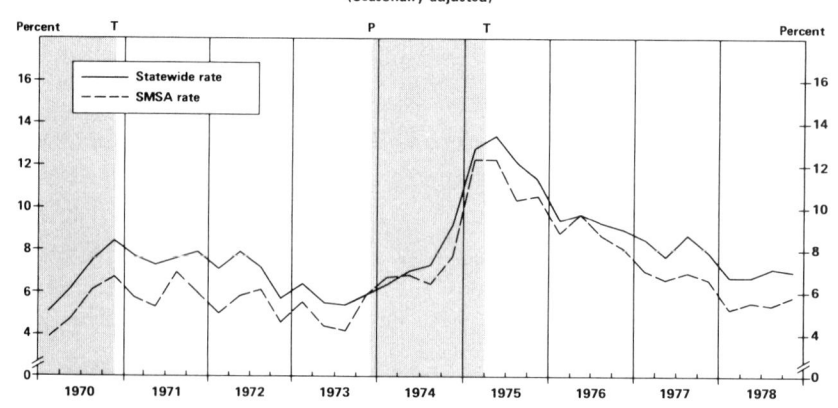

Source: Michigan Employment Security Commission.
Note: Seasonal adjustment by the W.E. Upjohn Institute.
Shaded areas indicate national recession periods as defined by the National Bureau of Economic Research, Inc.
P = peak and T = trough.

In contrast, government employment rose throughout the recession, as it did in almost all Michigan metropolitan areas.[1]

Employment in the Saginaw SMSA has expanded substantially since the last recession. By the end of 1978 manufacturing employment, which suffered the largest setback during the recession, was up 34.5 percent over its recession low, and as a result was 6.8 percent above its pre-recession peak. Only in the Ann Arbor-Ypsilanti and Bay City SMSAs had manufacturing employment risen farther above the pre-recession high. Combined with the steady increases in nonmanufacturing and government employment, the recovery in manufacturing resulted in a level of total wage and salary employment at the end of 1978 that was 13.8 percent above its previous cyclical high, which occurred in the third quarter of 1973.

The unemployment rate in the Saginaw SMSA remained below the state average throughout most of the 1970-78 period, although both exhibited similar cyclical patterns. Chart S2 shows that during the expansion period from 1971 through 1973 there was considerable difference between the state and local jobless rates. In the Saginaw SMSA the unemployment rate hit a seasonally adjusted low of 4.2 percent in the third quarter of 1973, compared to a statewide rate of 5.4 percent. Only in the Jackson SMSA was the unemployment rate lower. During the recession the unemployment rate rose considerably, hitting a high of 12.3 percent in the first and second quarters of 1975. At that time the jobless rate statewide was more than one percentage point higher.

As Chart S2 shows, the unemployment rate has fallen considerably since the recession as a result of the general business expansion nationwide and the recovery locally of the automobile industry and manufacturing employment. But, like other metropolitan areas in Michigan, the unemployment rate in the Saginaw SMSA at the end of 1978 was still above its 1973 low. For the entire nine-year period, this area's jobless rate drifted upward at an annual rate of 3.9 percent, compared to a more modest 2.8 percent trend statewide (see Table S2). The uptrend in the Saginaw SMSA ranked third behind the Jackson and Battle Creek SMSAs.

Average weekly initial claims for unemployment insurance are shown in Chart S3. The Saginaw SMSA experienced sizable

---

1. The exceptions are the Battle Creek, Flint, and Jackson SMSAs, where cyclical declines in government employment did occur during the recession.

increases in initial claims during both recessions, followed by declines during the business expansion periods. During the 1969-70 recession, average weekly initial claims in this area rose 456.1 percent, second only to the increase in the Flint SMSA and much more severe than the 193.2 percent rise statewide. The local upswing during the more severe recession of 1973-75 was 503.9 percent, again considerably above the statewide increase of 301.1 percent. During that recession, the rise in average weekly initial claims in the Saginaw SMSA ranked third behind the increases in the Ann Arbor-Ypsilanti and Flint SMSAs.

As Chart S3 shows, average weekly initial claims for unemployment insurance in this area declined considerably during the recent expansion period. In fact, in the third quarter of 1977, average weekly initial claims were below their pre-recession low. Thus, the Saginaw SMSA is the only metropolitan area in which initial claims have recently fallen below pre-recession lows. As Chart S3 shows, average weekly initial claims statewide have remained well above their 1973 low.

Despite the relatively good performance in the last three years, average weekly initial claims rose at an annual rate of 3.8 percent over the entire 1968-78 period. That is not only lower than the 7.3 percent growth rate for the state as a whole, it also represents one of the lowest among the state's 11 metropolitan areas. The Bay City and Flint SMSAs did experience lower growths than the Saginaw SMSA, with Flint registering the lowest rate of increase, 1.6 percent, over the eleven-year period.

Finally, the average workweek of manufacturing production workers in the Saginaw SMSA exhibits a high degree of cyclical sensitivity, similar to other areas that are heavily dependent on the automobile industry. During the 1968-78 period shown in Chart S4, the average workweek fluctuated within a range extending from a low of 37.7 hours in the third quarter of 1971 to a high of 47 hours in the fourth quarter of 1977. Only the Flint SMSA had a comparable range of variation in hours worked in the local manufacturing sector.

In the Saginaw SMSA the average workweek actually declined more during the 1969-70 recession than during the 1973-75 slump, 14.3 percent compared to 12.5 percent. The latter exceeded the 9.7 percent fall for the state as a whole and ranked fourth behind declines in the Ann Arbor-Ypsilanti, Flint, and Lansing-East Lansing SMSAs. Chart S4 shows that the expansion in the average

Saginaw SMSA

CHART S 3
AVERAGE WEEKLY INITIAL CLAIMS FOR UNEMPLOYMENT INSURANCE, STATE PROGRAMS
SAGINAW SMSA
(Seasonally adjusted)

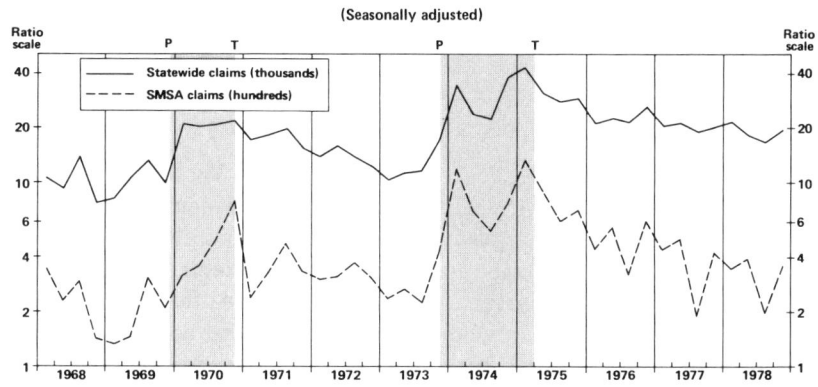

Source: The W.E. Upjohn Institute, based on data from the Michigan Employment Security Commission and the U.S. Department of Labor.

Shaded areas indicate national recession periods as defined by the National Bureau of Economic Research, Inc.
P = peak and T = trough.

CHART S 4
AVERAGE WEEKLY HOURS OF PRODUCTION WORKERS IN MANUFACTURING INDUSTRIES
SAGINAW SMSA
(Seasonally adjusted)

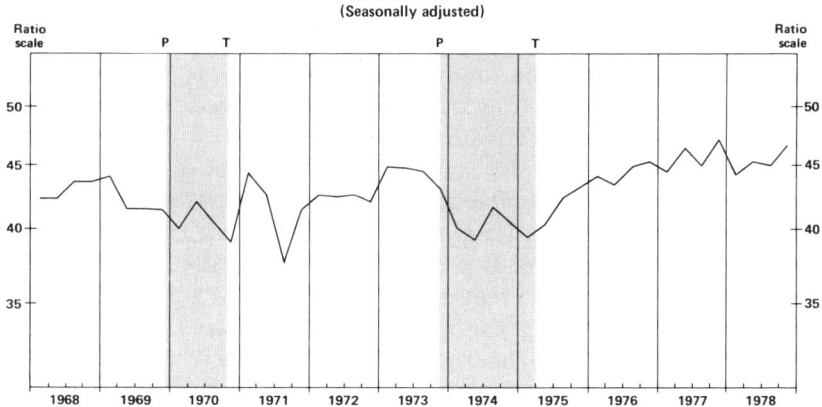

Source: Michigan Employment Security Commission.
Note: Seasonal adjustment by the W.E. Upjohn Institute.

Shaded areas indicate national recession periods as defined by the National Bureau of Economic Research, Inc.
P = peak and T = trough.

workweek has been sizable since the recession ended. In fact, the most recent high of 47 hours in the fourth quarter of 1977 was well above the 39.2 hours recorded during the second quarter of 1974, and exceeded by slightly more than 2 hours the pre-recession high of 44.8 hours.

Construction

As is the case in other metropolitan areas in Michigan, building activity in the Saginaw SMSA exhibits a high degree of cyclical volatility. New building permits for private housing are shown in Chart S5. The unadjusted quarterly index (dashed line) varies considerably from quarter to quarter. These fluctuations are reduced and the cyclical swings revealed by a moving average of the unadjusted data (solid line). As is the case in all Michigan metropolitan areas, the decline during the 1969-70 recession was much less severe than that which took place during the 1973-75 recession. In the former period, the moving average fell 31.5 percent, and during the more recent recession it dropped a sizable 73.2 percent. Both exceeded the declines registered statewide and nationwide. It should be noted, however, that neither of the downswings in the Saginaw SMSA were the most severe among the state's metropolitan areas. In fact, the contraction in the late 1960s was one of the mildest among Michigan metropolitan areas. During the 1973-75 recession the slump in the Saginaw SMSA ranked fifth behind relatively large downswings in the Ann Arbor-Ypsilanti, Battle Creek, Flint, and Kalamazoo-Portage SMSAs. Thus, over the period shown in Chart S5, new building permits in the Saginaw SMSA were much more volatile than for the state as a whole, but among the 11 metropolitan areas in Michigan their cyclical swings were comparatively mild. It should also be noted that, like all other Michigan metropolitan areas except Jackson, the upswing in new building permits during the recent expansion has failed to reach levels attained before the recession. Preliminary data for 1978 indicate that the upward momentum in this series may very well have ended.

In contrast to new building permits, construction employment in this area experienced a severe contraction during the last recession (see Chart S6). From peak to trough it fell 40.5 percent, the largest relative decline among Michigan metropolitan areas, and well in excess of the 23.5 percent drop statewide. Chart S6 shows, however, that employment in the construction industry has recovered to some extent since the recession low in the second quarter of 1975.

Saginaw SMSA

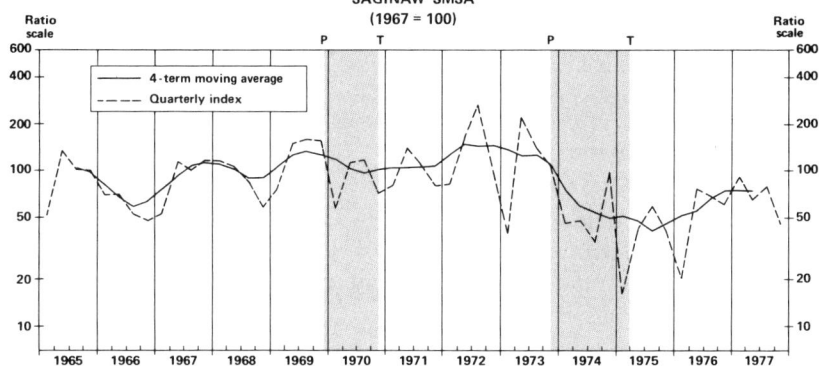

CHART S 5
INDEX OF NEW BUILDING PERMITS, PRIVATE HOUSING
SAGINAW SMSA
(1967 = 100)

Source: The W.E. Upjohn Institute. Index based on the U.S. Department of Commerce, Bureau of the Census, *Construction Reports — Housing Authorized by Building Permits and Public Contracts, C-40.*
Shaded areas indicate national recession periods as defined by the National Bureau of Economic Research, Inc.
P = peak and T = trough.

CHART S 6
CONSTRUCTION EMPLOYMENT, SAGINAW SMSA
(Seasonally adjusted)

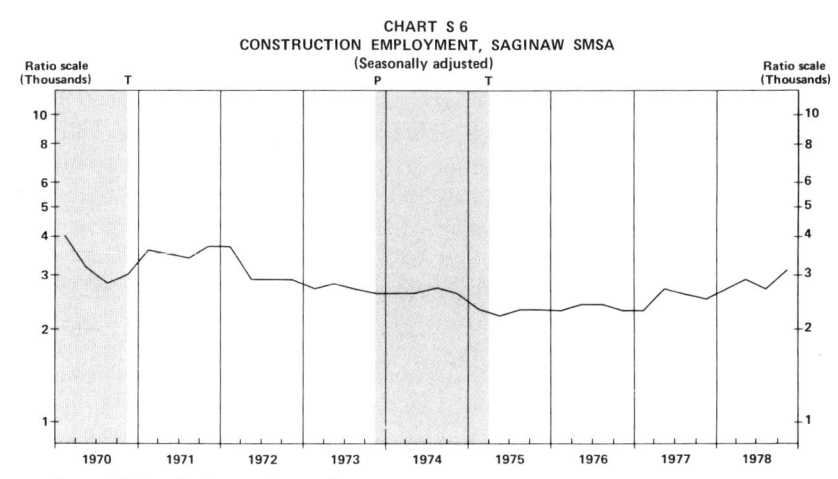

Source: Michigan Employment Security Commission.
Notes: Seasonal adjustment by the W.E. Upjohn Institute.
Shaded areas indicate national recession periods as defined by the National Bureau of Economic Research, Inc.
P = peak and T = trough.

In the fourth quarter of 1978 construction employment was up about 41 percent over its recession low. However, it was still 16 percent below the pre-recession high of the first quarter of 1972. Over the entire period from 1970 to 1978 construction employment has declined at an annual rate of 3.7 percent, which is the steepest downtrend among the 11 metropolitan areas in the state. Thus, the Saginaw SMSA is one of the six metropolitan areas in Michigan where employment in the local construction industry has experienced negative growth since 1970.

TABLE S3

Average Annual Growth Rates of Selected Commercial Banking Indicators Saginaw SMSA and Michigan[a]

(percent)

Indicator	Saginaw SMSA	Michigan
Demand deposits (current dollars)	6.3	3.7
Deflated demand deposits[b]	-0.6	-3.0
Total deposits (current dollars)	10.3	7.3
Deflated total deposits[b]	3.2	0.4
Total loans (current dollars)	11.6	7.6
Commercial and industrial loans (current dollars)	14.3	7.3
Consumer installment loans (current dollars)	14.5	9.1

[a] Except where indicated otherwise, estimated growth rates are based on log-linear trends for the 1970-78 period.

[b] Current-dollar values adjusted for changes in the U.S. Consumer Price Index.

Banking Activity

Loan and deposit growth at commercial banks in the Saginaw SMSA was considerably higher than that of the state over the 1970-77 period. Table S3 shows that current-dollar demand deposits and total deposits grew at annual rates of 6.3 percent and 10.3 percent, respectively, over this eight-year period. The comparable growth rates statewide were 3.7 percent for demand deposits and 7.3 percent for total deposits. The growth in the current-dollar

Saginaw SMSA 181

CHART S 7
INDEX OF DEFLATED TOTAL DEPOSITS AND INDEX OF DEFLATED DEMAND DEPOSITS
(1972 = 100)
(Seasonally adjusted)

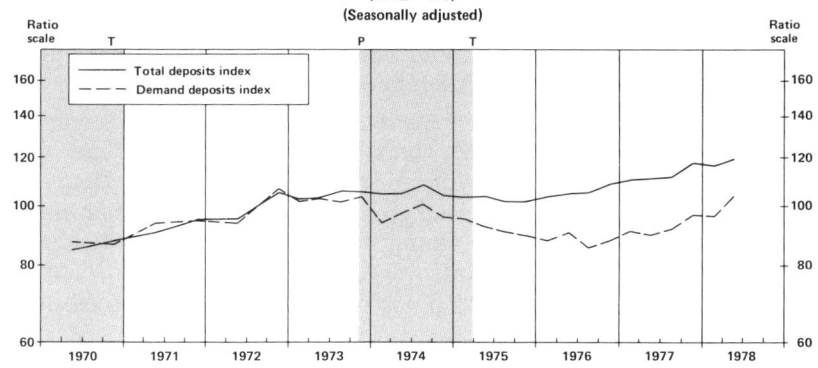

Source: The W.E. Upjohn Institute. Indices are based on data from the Federal Reserve Bank of Chicago.
Note: Shaded areas indicate national recession periods as defined by the National Bureau of Economic Research, Inc. P = peak and T = trough.

CHART S 8
COMMERCIAL BANK LOANS, SAGINAW SMSA
(Current dollars)

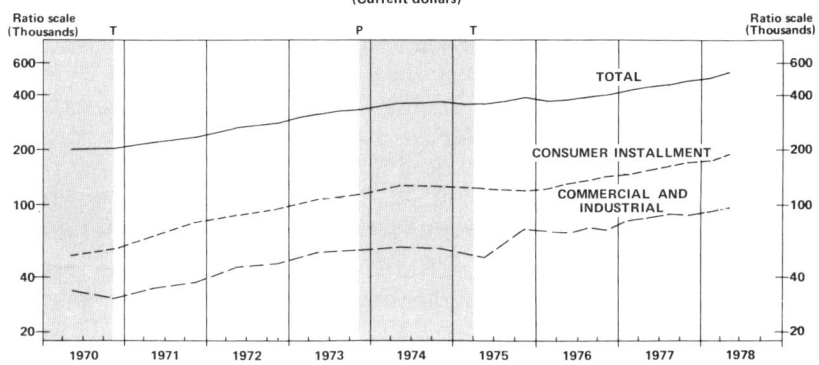

Source: Federal Reserve Bank of Chicago.
Notes: Seasonal adjustment of total loans by the W.E. Upjohn Institute. Other loans are not seasonally adjusted.
Shaded areas indicate national recession periods as defined by the National Bureau of Economic Research, Inc. P = peak and T = trough.

volume of demand deposits in the Saginaw SMSA ranked second among the state's 11 metropolitan areas, just behind the 6.9 percent growth in the Kalamazoo-Portage SMSA. In addition, this area's growth rate in total deposits was the highest, just as the annual growth rates for the current-dollar volume of loans, shown in Table S3, were the highest among Michigan metropolitan areas.

The strong uptrend in the current-dollar volume of deposits bolstered the deflated value of the two series shown in Chart S7. Although the cyclical downswing in the Index of Deflated Demand Deposits was long and amounted to a 19.9 percent drop from peak to trough, it was the mildest slump in this index among Michigan metropolitan areas. That was also the case for the Index of Deflated Total Deposits, which fell just 6 percent during the last recession.

Chart S7 shows that both indexes have moved up considerably during the recent expansion period. By mid-1978 the total deposits index had increased 17.1 percent from its recession low and stood 10.1 percent above its pre-recession peak. Although the demand deposits index had expanded 21.6 percent by mid-1978, it was still slightly below its pre-recession high, a situation similar to other metropolitan areas in the state.

Although there was a slight falloff in the current-dollar volume of commercial and industrial loans and consumer installment loans during the recession, Chart S8 shows that total loans grew steadily throughout the 1970-78 period, essentially uninterrupted by the national recession. Thus, the Saginaw SMSA joins the Ann Arbor-Ypsilanti and Muskegon SMSAs as areas in which the dollar volume of total loans did not decline during the 1973-75 slump in overall business activity. By mid-1978 the loan categories shown in Chart S8 were considerably above their pre-recession levels.

In summary, the Saginaw SMSA exhibited relatively rapid growth and a comparatively higher degree of stability in its banking activity than other metropolitan areas in Michigan. Local employment also experienced fairly strong growth, but was characterized by sizable cyclical swings in the manufacturing sector, not an unexpected result of the heavy concentration on automobile-related production in the area. Therefore, like other metropolitan areas in the state which are heavily dependent on the automobile industry, the Saginaw SMSA experienced large declines in manufacturing employment during recessions. But these slumps were followed by vigorous expansions. On the other hand, the behavior of construction employment, which also suffered a sizable downswing during the last recession, over the 1970-78 period exhibited the largest negative growth rate among the 11 metropolitan areas in Michigan.

Dr. Kozlowski prepared this study while associated with the Upjohn Institute as a senior staff economist. He was assisted by Phyllis Buskirk, a senior staff research associate with the Institute. Kozlowski is currently an assistant professor of business economics in the College of Business Administration, The University of Toledo.

HC
107
.M5
K69
1979

BH
12/10/84

#120863

$10.59

Kozlowski, Paul J.
  Business conditions in Michigan
metropolitan areas / by Paul J.
Kozlowski, with assistance from Phyllis
R. Buskirk. -- Kalamazoo, Mich. : W.E.
Upjohn Institute for Employment
Research, c1979.
  vii, 182 p. : ill. ; 23 cm.
  "December 1979."
  Includes bibliographical references.
  ISBN 0-911558-71-3 (pbk.)
  1. Michigan--Economic conditions.
2. Metropolitan areas--Michigan.
I. Buskirk, Phyllis R., jt. author.
II. W. E. Upjohn Institute for
Employment Research. III. Title

MiRoscK19 DEC 84   5726472   QZTAx1   79-24777